Selected Titles in This Series

One Field, Many Paths:
U. S. Doctoral Programs
in Mathematics Education

CBMS

Conference Board of the Mathematical Sciences

Issues in Mathematics Education

Volume 9

One Field, Many Paths:
U. S. Doctoral Programs
in Mathematics Education

Robert E. Reys
Jeremy Kilpatrick
Editors

American Mathematical Society
Providence, Rhode Island
in cooperation with
Mathematical Association of America
Washington, D. C.

One Field, Many Paths: U. S. Doctoral Programs in Mathematics Education was made possible by the National Science Foundation (ESI-9810644). The ideas and opinions expressed by the papers in this volume are those of the authors and editors, consequently they do not necessarily reflect the position of the National Science Foundation.

2000 *Mathematics Subject Classification.* Primary 97–06.

Library of Congress Cataloging-in-Publication Data

One field, many paths : U.S. doctoral programs in mathematics education / Robert E. Reys, Jeremy Kilpatrick, editors.
    p. cm. — (Issues in mathematics education, ISSN 1047-398X ; v. 9)
    Includes bibliographical references
    ISBN 0-8218-2771-5 (alk. paper)
    1. Mathematics—Study and teaching (Graduate)—United States—Congresses. 2. Mathematics teachers—Training of—United States—Congresses. I. Reys, Robert E. II. Kilpatrick, Jeremy. III. Series.

QA13 .O54 2001
510′.71′173—dc21
                                                          2001018843

CONTENTS

PART 3: RELATED ISSUES

PART 4: REACTIONS AND REFLECTIONS

PART 5: IDEAS FOR ACTION

PREFACE

The first doctorates in mathematics education in the United States were awarded in the early years of the twentieth century. By the end of the century, more than 100 U.S. institutions were offering such doctorates, and between 50 and 115 doctorates were being awarded annually with mathematics education as their major area. In some instances, the institution offered a specific doctoral program in mathematics education, but in many others, the program was more general—such as general education, curriculum and instruction, or mathematics—and mathematics education might be only one of several possible specializations within it. Many of the mathematics education doctoral programs were located in departments of mathematics—particularly in institutions that had begun as teachers colleges—but the great majority were to be found in schools or colleges of education.

As the century ended, employment opportunities in mathematics education were expanding, and there was a severe shortage of qualified applicants for new faculty positions. University mathematics departments were seeking mathematics educators not only to teach courses in mathematics or mathematics education for teachers but also to conduct research into the teaching and learning of undergraduate mathematics. School districts and state departments of education were seeking mathematics educators for their staffs who could lead projects in curriculum development, professional development, and assessment. The question of what preparation a doctoral program in mathematics education could and should provide candidates for these new positions was becoming increasingly knotty.

As a field that straddles the liberal arts and the professions, mathematics education has always had a blurred institutional identity. It is not surprising, therefore, that doctoral studies in the field take many forms. Although from time to time, particularly in the last three decades, surveys have been conducted of U.S. doctoral programs in mathematics education, there has been little in the way of up-to-date, substantive information on those programs, and no information on proposed new programs or planned changes in programs.

To begin the process of examining, discussing, and making recommendations for doctoral programs in mathematics education, a National Conference on Doctoral Programs in Mathematics Education was held in October 1999 at Lake Ozark, Missouri. It was funded by a grant from the National Science Foundation and was preceded by a year of preparation that included a comprehensive survey of programs.

Both the survey and the conference were designed and conducted under the direction of an organizing committee. The committee met in Atlanta in October 1998 to provide ideas for the survey and suggestions for the conference themes and speakers. Members of the committee were F. Joe Crosswhite, The Ohio State University (emeritus); Elizabeth Fennema, University of Wisconsin (emerita); Joan Ferrini-Mundy, Michigan State University; Martin Johnson, University of Maryland; Jeremy Kilpatrick, University of Georgia; Mary Lindquist, Columbus State University; Pat Thompson, Vanderbilt University; and James Wilson, University of Georgia. The survey was conducted in the spring of 1999 by Robert Reys, Bob Glasgow, Gay Ragan, and Ken Simms, all at the University of Missouri.

The conference was designed to provide a dialogue regarding

- the nature of current doctoral programs in mathematics education,
- ways of strengthening such programs, and
- suggestions and guidelines for faculty engaged in restructuring an existing program or creating a new one.

The conference was by invitation only. Invitations were sent to one faculty member from each of the institutions identified by the survey as among the 30 largest U.S. producers of doctorates in mathematics education. Representatives from 43 institutions participated, and these institutions accounted for about 60 percent of the doctorates in mathematics education awarded during the last 20 years. Invitations were also extended to faculty members involved in establishing recent doctoral programs in mathematics education at their institutions; to individuals with perspectives on non-U.S. programs; and to others representing organizations such as the National Council of Teachers of Mathematics, the National Council of Supervisors of Mathematics, the Mathematical Association of America, the American Mathematical Association of Two-Year Colleges, the Association of Mathematics Teacher Educators, and the National Science Foundation. A complete list of participants is provided in Appendix A.

Before the conference, the participants were sent a draft of the survey results. One conference session was devoted to a discussion of those results and their implications. Other conference sessions focused on the genesis of U.S. doctoral programs, major program components, the match between preparation and employment, challenges in developing new and revising existing programs, doctoral programs from an international perspective, and actions to be taken as a consequence of the conference. Groups of participants discussed preparation in research, in mathematics, in mathematics education, and in teaching. Discussions were also held on the changing nature of dissertations, program features beyond course work, distance learning, and recruiting and funding doctoral students. In addition to scheduled presentations, the conference was organized to encourage and facilitate informal discussions.

One Field, Many Paths: U.S. Doctoral Programs in Mathematics Education, the first organized collection to focus specifically on doctoral programs in mathematics education, contains papers prepared for the conference as well as papers submitted afterward. The papers have been reviewed, edited, and organized into five sections dealing with background information on programs, core components of programs, issues, reactions and reflections, and ideas for future action. The "Background" provides different perspectives of doctoral programs in mathematics education. The papers in "Core Components" highlight elements that are common in most doctoral programs, including course work as well as experiences that go beyond courses. A range of other issues surfaced during the conference, such as recruitment and organizing new programs, and these issues constitute part three, "Related Issues." Some thoughts from participants after the conference are included in "Reactions and Reflections." The final section, "Ideas for Action," includes a single paper that provides a brief synthesis of the conference and offers suggestions for future action.

The "Background" includes the initial keynote presentation by Eileen Donoghue that highlighted the evolution of doctoral programs in the United States. This is followed by a summary of survey results by Robert Reys, Bob Glasgow, Gay Ragan and Ken Simms that provides information about institutions and doctoral graduates since 1980, as well as some data on faculty and program requirements. Mathematics educators (Diane Briars, Terry Crites, Skip Fennell, Susan Gay, and Harry Tunis) whose careers have taken them in different directions reflect on the match between their current position and their doctoral preparation. An international perspective of doctoral preparation in mathematics education by Alan Bishop (Australia) serves as a reminder that the model for doctoral programs in the United States is not reflected in many other countries.

Jim Fey uses broad strokes to highlight core elements of doctoral programs and sets the stage for the remaining papers within the "Core Components." Four core areas are addressed, including mathematics content (John Dossey and Glenda Lappan); research (Tom Carpenter and Frank Lester); mathematics education (Norma Presmeg and Sigrid Wagner); and teaching (Diana Lambdin and Jim Wilson). Research leading to a doctoral dissertation is a common element of every doctoral program. Lee Stiff highlights some ways that dissertations are changing. Quality doctoral programs provide opportunities for professional growth and development beyond course work, and some of these options are highlighted by Glen Blume.

Additional issues surfaced during the conference, and these are addressed in "Related Issues." For example, some institutions (East Carolina University, Portland State University, San Diego State University, Illinois State University and Montclair State University) are initiating new doctoral programs in mathematics, and faculty from those institutions (Robert Hunting, Mike Shaughnessy, Judy Sowder, Carol Thornton, Ken Wolff) share some problems and challenges associated with establishing new programs. Other institutions (Oklahoma State University, Stanford University, Rutgers University, University of Mississippi and George Mason University) are revamping their existing programs; Doug Aichele, Jo Boaler, Carolyn Maher, David Rock and Mark Spikell highlight ways these programs are evolving. Recruiting and supporting doctoral students is a common concern for all programs, and approaches used in different institutions are summarized by Ken Wolff. Program outreach to serve non-resident students is another major issue, and distance learning has the potential to address this need. Charles Lamb discusses distance learning and provides some suggestions. Richard Lesh, Janel Crider and Edith Gummer propose a model for collaborative doctoral programs in mathematics education that involve multiple campuses as well as distance learning.

"Reactions and Reflections" includes participants' thoughts on different issues, including policy (Vena Long), challenges of small doctoral programs in mathematics education (Jenny Bay), implications for new programs (Alfinio Flores), why an international student chose to enter a doctoral program in the U.S. (Thomas Lingefjärd), and personal reflections from a doctoral student participating in the conference (Gay Ragan).

At the closing session, participants identified several prevalent themes that surfaced during the conference. There was clear consensus on the need for more information about doctoral programs in mathematics education. A more controversial issue related to the need for and value of guidelines for such programs. Although no consensus was reached, there was agreement that this issue was important and needed further deliberation. The paper in the "Ideas for Action" by Jim Hiebert, Jeremy Kilpatrick and Mary Lindquist reflects on the conference discussions, and offers ideas for action that will continue the dialogues established at the conference and lead to the improved quality of doctorates in mathematics education.

To all the conference participants, we express our gratitude for the many thoughtful ideas and suggestions that were proposed. This publication allows some of that information to be communicated to a broader audience. We are especially grateful, therefore, to those participants who contributed papers in this volume. We recognize the many demands on their time and appreciate their willingness to share their knowledge and experience. A special thanks to Bob Glasgow, Gay Ragan and Tim Sanders (doctoral students at that time) for their willingness to assume a range of tasks and duties that contributed to the success of the conference, and their contributions to the development of this volume, and to Sara Priddy for graphic design.

Finally, we thank the National Science Foundation for providing the financial support for the project and the Conference Board of the Mathematical Sciences for making the conference papers available through this publication. We hope *One Field, Many Paths: U.S. Doctoral Programs in Mathematics Education* will stimulate discussion and prove useful in preparing future recipients of doctorates in mathematics education.

Robert E. Reys
121 Townsend Hall
University of Missouri
Columbia, MO 65211
reysr@missouri.edu

Jeremy Kilpatrick
105 Aderhold Hall
University of Georgia
Athens, GA 30602-7124
jkilpat@coe.uga.edu

PART 1: BACKGROUND

CBMS Issues in Mathematics Education
Volume 9, 2001

MATHEMATICS EDUCATION IN THE UNITED STATES: ORIGINS OF THE FIELD AND THE DEVELOPMENT OF EARLY GRADUATE PROGRAMS

Eileen F. Donoghue, College of Staten Island, City University of New York

This paper examines two closely related topics: the origins of mathematics education as a field of study in the United States and the development of early graduate programs in the field. It focuses upon events occurring between 1892 and 1912, when the foundations for the field of mathematics education were established.

The first section of the paper explains how the field emerged at the end of the nineteenth century and gained recognition during the early years of the twentieth century. It also identifies national and international events that influenced the establishment of the field. The second section examines the development of graduate programs with particular attention to the first two doctoral programs in mathematics education. What were the components of these early programs? Who established them, who enrolled in them, and why did they do so?

This volume concerns the current state and anticipated future needs of doctoral study in mathematics education. Yet, there is a place for the past in such discussions. Reflection upon issues that confronted the founders offers insights and possibly answers to questions faced by those in the field today. We may well find inspiration in the dedication and vision of those who pioneered the field at the turn of the last century.

ORIGINS OF MATHEMATICS EDUCATION AS A FIELD OF STUDY IN THE UNITED STATES

The field of mathematics education in the United States traces its origins to the 1890s. During the early 1900s, the foundation was laid for graduate study in the emerging field. By 1912, mathematics education was recognized as an area of study distinct from, but intimately tied to, the field of mathematics itself (Donoghue, in press; see also Jones & Coxford, 1970). To understand how and why this occurred, it is necessary first to survey briefly the 15-year period from 1875 to 1890, particularly developments in three areas: the schools, mathematics, and higher education.

PRELUDE, 1875-1890

The period 1875 to 1890 in the United States was an era of great growth in the cities. The nation's industrial base expanded, prompting a migration of workers from rural farms to urban factories. Immigration, mainly from Europe, added to city populations. American high schools also developed during this period. The number of students attending these schools was not large, but it was growing.

Mathematics was viewed as a "tool" subject, not a discipline to be studied for its own merits. It was valued for its practical applications in commerce, science, and engineering, or for its presumed usefulness in developing logical thinking and mental discipline—good habits of mind. With the exception of Johns Hopkins University, few U.S. schools offered students the opportunity to study advanced mathematics. Those who wished to do serious theoretical work at the doctoral level had to travel abroad. Most sought out the University of Göttingen, where leading mathematicians such as Karl Weierstrass and Felix Klein conducted seminars on research topics. Upon returning to the United States, these scholars took positions on college or university faculties and joined with colleagues in other disciplines who had had similar experiences abroad to call for more advanced programs of study at American institutions that aspired to true university status. Their efforts contributed to a transformation of American higher education heralded by the rise of the university during the 1870s and 1880s. The first American university with true graduate-level study as its mission was Johns Hopkins, founded in 1876. James Joseph Sylvester was recruited from England to establish a doctoral program in mathematics that provided the kind of research guidance in theoretical topics that previously had not been available in the United States. When Sylvester departed in 1883, the university was unsuccessful in securing another mathematician of comparable stature (see Rudolph, 1962; Smith & Ginsburg, 1934).

As universities reformed themselves, colleges and normal schools began to examine their own missions, particularly with regard to the preparation of teachers for the growing number of high schools. Traditionally, the normal schools concentrated on pedagogy and trained teachers in one- or two-year programs for service in elementary schools (grades 1-8). Colleges provided "academic" courses in specific subject areas but no "professional" preparation for the job of teaching in secondary schools and academies. As more public high schools opened, questions arose regarding the appropriate preparation for high school teachers. Normal schools insisted that training in subject matter alone was inadequate for the realities of classroom teaching. Colleges maintained that the pedagogical focus of the normal school did not provide teachers with sufficient depth of content knowledge to teach specific subjects or to prepare secondary school students for college-level work (see Brown, 1911; Gibb Karnes & Wren, 1970; Luckey, 1903).

The controversy spurred adaptations by each type of institution. By 1890, many normal schools had begun to offer more rigorous academic courses and many colleges and universities had developed courses in pedagogy. The college pedagogy courses usually involved general methods and sometimes were introduced as "applied" psychology. At the time, psychology was viewed as a branch of philosophy, thus, many pedagogy courses were first offered within philosophy departments. As enrollments in pedagogy courses increased, separate programs or departments of pedagogy were established in the colleges and universities. This led to further specialization at some institutions that offered methods courses specific to an academic discipline such as English or science; however, in 1890 no special methods courses for teaching secondary school mathematics existed in any normal school, college, or university in the United States (*Report of the American Commissioners of the International Commission on the Teaching of Mathematics*, 1912). Soon, this would change.

ORIGINS, 1890-1900

During the 1890s, significant events occurred in the schools, in mathematics, and in higher education. Developments in one sphere influenced those in the other two; together, these developments led to formation of the fledgling field of mathematics education (see Jones & Coxford, 1970; Osborne & Crosswhite, 1970; Stanic, 1986, 1987).

Schools

The years 1890 to 1900 witnessed rapid growth in American schools. Enrollment in public high schools more than doubled from 202,963 to 519,251. (U.S. Bureau of the Census cited in Tyack, 1967, 468-469). Increasingly, the high schools were viewed as the "people's college."

In 1893, the Committee on Secondary School Studies (Committee of Ten), appointed by the National Educational Association (1893) and chaired by Charles Eliot of Harvard University, issued its report on the nation's schools. The report recommended curricular reforms with special emphasis on needed changes in secondary school courses intended to prepare students for college study. A conference committee was appointed to formulate recommendations for mathematics. The conference members consisted of ten men, almost all from colleges or universities, led by Simon Newcomb, an applied mathematician and astronomer who headed the Nautical Almanac Office in Washington, D.C., and had a part-time appointment to the faculty of Johns Hopkins.

For the elementary school level, the Mathematics Conference Report (National Educational Association,1893/1970b) recommended the introduction of concrete geometry and radical changes in the arithmetic curriculum to eliminate obsolete topics. For the secondary school, it called for the coordination of advanced algebra and geometry courses so that the two could be taught simultaneously, but separately, in the second and third years of high school. The report of the Committee of Ten received widespread attention during the 1890s and early 1900s.

Mathematics

The American mathematics research community was established during the 1890s (see Parshall & Rowe, 1994). Three events that contributed to a coalescence of the community were the founding of the University of Chicago, the opening of the World's Columbian Exposition, and the expansion of the New York Mathematical Society.

In 1892, the private, coeducational University of Chicago opened, led by William Rainey Harper and financed by John D. Rockefeller. To form the mathematics department, Harper brought Eliakim Hastings Moore from Northwestern University in nearby Evanston to the new campus on the south side of Chicago. With Harper's support, Moore offered positions to individuals who shared his commitment to developing a strong research program for graduate students. Two such individuals were German mathematicians Oskar Bolza and Heinrich Maschke, who had studied under Felix Klein. By the end of the decade, Chicago had become the leading department for mathematical study and research in the United States.

As the Committee of Ten issued its report on the nation's schools in 1893, the World's Columbian Exposition opened in Chicago on land adjacent to the new university. In conjunction with the exposition, the American mathematics community hosted an International Mathematical Congress. Felix Klein (1893), the eminent German mathematician, accepted an invitation to address the congress on behalf of his government and brought with him papers written by other German luminaries. The American mathematicians who attended the congress were impressed by Klein's insights into the progress and problems of mathematics during the nineteenth century. Klein

(1893/1911) agreed to conduct a two-week series of lectures at Northwestern University after the close of the congress. This series, known as the Evanston Colloquium, was a signal event in the emergence of a community of research mathematicians in the United States. Notably, Klein concluded the colloquium with a lecture on teaching and learning mathematics at Göttingen University. He described the general course of study for those who intended to become mathematics or physics teachers in Germany's secondary schools, and he related some of his own teaching methods. Klein's gestures suggested a collegial respect for the young American mathematics community that bolstered their confidence.

The following year, 1894, the New York Mathematical Society changed its name to the American Mathematical Society (AMS), a change that reflected the growth of its membership. The society had been founded at Columbia University on Thanksgiving Day, 1888. Thomas Fiske, a graduate student who would later join the Columbia faculty, convinced five other members of the mathematics department to join him in organizing the group. By 1891, membership had grown to over 200, but just 20% resided in the New York City area. It was unlikely that members from distant states such as Alabama, Minnesota, and Wyoming could attend the New York meetings with any regularity; however, the *Bulletin* published by the society kept them informed of activities and discussions (Archibald, 1938; New York Mathematical Society, 1892; "Notes," 1892). In 1896, the strong mathematical community in Chicago formed its own section of AMS and gained further influence over the direction of the society. Indeed, in 1899 a committee formed by the Chicago Section and led by Jacob William Albert Young issued a report for the National Educational Association (1899/1970a). The report built upon the work of the Committee of Ten but broke new ground by including recommendations for the preparation of high school mathematics teachers. In part, these recommendations grew out of developments that had occurred in American higher education during the 1890s.

Thus, by 1900 the University of Chicago had become the American center of mathematical research and was producing strong doctoral graduates for university faculties; American mathematicians had entered the international arena through the International Mathematical Congress; and AMS had become an influential voice for the study of mathematics in the United States.

Higher Education

The field of mathematics education had its beginnings in the 1890s when normal schools and universities began to offer special methods courses for teaching mathematics at the secondary level. The University of Michigan, first to establish a permanent chair in the science and art of teaching in 1879, became the first institution to offer a course in mathematical pedagogy for the secondary school. In 1892, Wooster Woodruff Beman, senior professor of mathematics, introduced two Teachers' Seminars open to both undergraduate and graduate students. One seminar considered the teaching of algebra and the other the teaching of geometry (University of Michigan, 1892-1893).

The following year, Michigan State Normal School (now Eastern Michigan University) in neighboring Ypsilanti became the second institution to offer a separate course in the teaching of mathematics. David Eugene Smith, who had been recruited to strengthen the mathematics program at the school, inaugurated an elective course on methods in algebra and geometry. The course dealt with issues in teaching, but it also included study of the history of these two subjects and how they had been taught (Michigan State Normal School, 1894-1895).

In the 1895 summer session, J. W. A. Young, a mathematics faculty member at the University of Chicago, offered a course on mathematical pedagogy. The course examined the teaching of both preparatory and collegiate mathematics (University of Chicago, 1894-1895). The introduction of this course into the set of offerings by the most important mathematics department in the country served as an endorsement of study in this new area of mathematical pedagogy.

Teachers College introduced the course Methods of Teaching Mathematics in the fall semester of 1895. The course focused upon the secondary school and was taught by Charles Earl Bicklé with assistance from Julia Helen Wohlfarth (Teachers College, 1895-1896). At the time, Teachers College was essentially a private normal school with loose ties to Columbia University. In less than a decade, Teachers College would transform itself into a center for graduate study of education more closely affiliated with Columbia.

In 1896, Syracuse University, a coeducational, church-affiliated school, became the fifth American institution to offer a course in mathematical pedagogy. Syracuse had decided to abandon the general methods approach for prospective secondary school teachers and to inaugurate subject-specific pedagogy courses within the respective academic departments. In mathematics, department head and university dean John French offered a one-semester teachers' course for those preparing to teach in a high school (Syracuse University, 1896-1897).

By 1900, four other institutions—University of Pennsylvania, Indiana University, Albany Normal College (now SUNY), and University of Illinois—had begun to offer courses in mathematical pedagogy for the high school. Over the following decade, twenty-five additional institutions would introduce similar courses (International Commission on the Teaching of Mathematics, 1911).

During the 1890s, then, American high schools experienced growing enrollments and felt the push for higher standards for students and teachers. The American mathematics community, though still small, grew in numbers, confidence, and stature. The nation's normal schools, colleges, and universities blended the academic and pedagogic aspects of their preparation programs for high school teachers and, as a result, offered the first American courses in mathematics education.

FOUNDATIONS, 1900-1912

The foundations of the field of mathematics education were established during the first decade of the twentieth century. Four factors contributed to this establishment: regional associations were formed; journals focused upon mathematics education were founded; international links were forged; and graduate programs were established. These factors paralleled factors that had contributed to the formation of the mathematics research community during the late 1800s.

Regional Associations

In his address as retiring president of AMS, Moore (1903/1926) recommended that the society welcome as members school teachers with a serious interest in mathematics. Subsequently, he urged mathematics teachers to form their own associations as well.

Moore believed that mathematics teachers at all levels should cooperate to form a network of associations across the country (Smith Papers, Moore to Smith, 12 August 1903).

In 1903, three regional and two state associations were formed. The New England Association of Teachers of Mathematics was organized by William Osgood of Harvard University and Harry W. Tyler of Massachusetts Institute of Technology. Thomas Fiske and David Eugene Smith convened a meeting at Teachers College, Columbia University that resulted in formation of the Association of Teachers of Mathematics in the Middle States and Maryland. In Moore's own city of Chicago, the Central Association of Physics Teachers agreed to extend membership to teachers of mathematics and the other sciences. The larger joint organization was renamed the Central Association of Science and Mathematics Teachers. State associations of mathematics teachers were formed in Washington and Kansas (see Central Association of Science and Mathematics Teachers, 1903; Central Association of Science and Mathematics Teachers, 1950; "The Meetings of the Association," 1904).

Journals

Two of the new associations founded journals that focused upon issues of teaching mathematics in the secondary school. The first such journal in the United States was the *Mathematical Supplement* to the central association's existing publication *School Science*. The *Supplement* first appeared in 1903. The following year, it became a separate journal with a new title, *School Mathematics*. By 1905, *School Mathematics* had merged with the science journal to form *School Science and Mathematics* ("Explanation," 1905; "Mathematical Supplement of School Science," 1903a, 1903b; Schreiber & Warner, 1950).

The Middle States and Maryland Association began to publish an annual Bulletin of its activities in 1904. In 1908, the association expanded publication to quarterly issues and renamed it the *Mathematics Teacher.*

International Links

A third factor that contributed to the emergence of mathematics education as a field was the international links forged during the 1890s and early 1900s. J. W. A. Young (1900) spent a year studying the Prussian system of education in mathematics and subsequently published a report of his findings. David Eugene Smith regularly traveled abroad to visit schools and universities. He arranged to attend the International Mathematical Congresses, quadrennial meetings instituted after the success of the Chicago congress. Smith's disappointment in the sessions identified by the congress as concerned with issues of teaching led to his proposal that an international group separate from, but affiliated with, the International Congress of Mathematicians (ICM) be established. The group's purpose would be to organize national committees that would report upon the status of mathematics teaching in their respective countries (see Smith, 1905).

As a result of Smith's efforts, the International Commission on the Teaching of Mathematics (ICTM) was established by the ICM at the 1908 Rome meeting. Felix Klein agreed to serve as the first president of the commission. He worked closely with Smith to guide the selection of national delegations. Each national delegation then organized its own system of committees to research and report upon the state of mathematics teaching in its schools and universities (see Smith Papers, ICTM Memoranda, n.d.).

At the 1912 international congress in Cambridge, the ICTM presented over 150 reports that provided for the first time a profile of mathematics education around the world (for example, see *Report of the American Commissioners of the International Commission on the Teaching of Mathematics*, 1912).

Graduate Programs

The fourth factor that contributed to the emergence of mathematics education as a field was the establishment of graduate programs that focused upon the teaching of mathematics. As in the case of theoretical mathematics in the 1890s, where a few strong departments set the pace for graduate and undergraduate study in the field, in mathematics education two institutions developed graduate programs that subsequently served as models: Teachers College of Columbia University and the University of Chicago. Through the efforts of David Eugene Smith at Teachers College and E. H. Moore, J. W. A. Young, and George Myers at the University of Chicago, these programs and their graduates helped to establish mathematics education as a field. It is to these two programs that we now turn our attention.

EARLY GRADUATE PROGRAMS IN MATHEMATICS EDUCATION

The first doctoral programs in mathematics education were developed at two universities, Columbia and Chicago, that had attached to themselves formerly independent teacher training institutions, Teachers College and the Chicago Institute, respectively. As constituents within a university system, the training institutions redefined their missions and developed into professional schools of education that emphasized graduate study and the preparation of educational leaders. In mathematics education, both Teachers College and the University of Chicago first established master's degree programs that incorporated investigation and thesis components. Doctoral programs followed within a few years. Although their orientations were different, both doctoral programs required substantial study of mathematics and scholarly research on a topic important for teaching or learning mathematics.

TEACHERS COLLEGE, COLUMBIA UNIVERSITY

When James Earl Russell arrived at Teachers College in 1898, he received a request from the board of trustees to prepare a plan for the future direction of the college. Russell believed that the college should redefine itself as a professional school of education comparable to Columbia's schools of law, medicine and applied science. The mission of the college would be to develop leaders in education, and the emphasis would shift toward graduate-level study. Impressed by Russell's novel idea, the trustees re-negotiated the terms of their agreement with Columbia along the lines of Russell's plan and then appointed Russell to the newly created position of dean. Russell moved decisively to lead the college in the direction he had outlined. He sought out the faculty to pioneer new areas of study and new degree programs. For mathematics Russell invited David Eugene Smith to join his effort (see Russell, 1900a, 1900b, 1937).

Characteristics of Smith's Program

Smith had led the innovative mathematics program at the Michigan State Normal School until 1898, when he accepted an administrative position as principal of Brockport (NY) Normal School. In 1900, Ginn and Company published Smith's seminal text for training teachers, *The Teaching of Elementary Mathematics*. The book grew out of Smith's fifteen years of experience preparing mathematics teachers for the elementary and secondary schools. By 1901, Smith had tired of the relentless duties of an administrator and wished to return to an academic role, where he could more readily pursue his scholarly interests in the history and teaching of mathematics (Russell Papers, David Eugene Smith to James Earl Russell, 16 December 1936). Thus, at age 41, Smith arrived in New York City to begin work at Teachers College.

The program Smith established at Teachers College retained characteristics of the programs he had developed at Cortland (NY) Normal School in the 1880s and Michigan State Normal School in the 1890s. But New York, with its turn-of-the-century energy and opportunity, and Teachers College, with Russell's vision for graduate-level study, provided an environment in which Smith's program could flourish.

Smith's programs shared three hallmarks: an emphasis on history, an international perspective, and a "dynamic" aspect. Study of the history of mathematics and of how mathematics had been taught were important components of every program Smith developed. In his view, knowledge of a discipline's history was a prerequisite to teaching it effectively. Smith insisted upon an international perspective because he believed that study of educational systems in different countries promoted a broader consideration of issues in teaching mathematics. In his travels, Smith himself had found that much could be learned from others. Finally, Smith held that neither mathematics nor its teaching was a static subject; consequently, mathematics education must continue to evolve. The teacher was to be an active contributor to that development. These were the principles that guided Smith and his students (see Donoghue, in press).

The particulars of course work were determined individually in consultation with Smith. Students arrived at the college with different academic credentials. Some held diplomas from normal schools; others had earned bachelor's or master's degrees. Those who wished to pursue graduate degrees usually were experienced teachers who sought advancement at secondary schools, normal schools, or colleges. As in the German universities Smith had visited, his program combined preparation for teaching at these different levels.

Master of Arts Degree

The master of arts degree required a minimum of one year of study, completion of eight semester-long courses, and submission of a report or essay that used "approved methods of critical investigation" (Teachers College, 1902, pp. 3-5). Typically, these reports were approximately 30 pages in length and included some form of original contribution to scholarship. Among the first graduates of Smith's master's program were John William Adams, Alice Mabel Gilliland, and Julia Elnora Richardson (Adams, 1902; Gilliland, 1903; Richardson, 1903).

Adams had earned a bachelor's degree from Cornell University in 1901 and enrolled immediately at Teachers College, where he received a master of arts degree in 1902. Adams's (1902) essay, "Correlation between Mathematics and Physics in American High Schools," addressed a topic that was championed by Moore and his colleagues at the University of Chicago and that generated considerable controversy in mathematics at the time. Although not a proponent of the correlation movement, Smith believed that study of such timely issues would prepare graduate students like Adams to take an active part as teachers in shaping their own profession.

Gilliland also attended Cornell, graduating in 1891. The following year she enrolled at the State Normal College in Albany for further training. Her 1903 master's essay at Teachers College, "Mathematics in the Great Public Schools of England," exemplifies another hallmark of Smith's program, the international perspective.

Richardson came to New York from the Middle West. She had received a diploma in 1892 from the State Normal School in Winona, Minnesota and had completed a B.S. degree at Northwestern University in 1899. Richardson's topic for her 1903 master's essay at Teachers College, "Influence of Pope Sylvester II (Gerbert) upon Mathematics about the Year One Thousand," reflected the third hallmark of Smith's program, an emphasis upon history. Richardson's sources included authoritative German- language histories of mathematics and biographies of Gerbert.

During the years 1900 to 1912, forty-four master's essays deposited in the Teachers College library dealt in some way with teaching mathematics. Among these essays, nine concerned the history of mathematics; nine examined curricula or texts; four reported upon educational systems in other countries and, in some cases, compared them with mathematics education in the United States; and four discussed recreational mathematics (see Teachers College, *Dictionary Catalogue*).

Two other master's essays of this period warrant mention: a 1907 report by Clifford Brewster Upton entitled "Modern Calculating Machinery and Its Bearing on the Teaching of Mathematics," and a 1912 report by M. West entitled "The Differential and Integral Calculus in Secondary Schools." These topics, more broadly construed, have been discussed and debated extensively during the last half of the twentieth century and have been examined in numerous research studies. Upton had been Smith's student at Michigan State Normal School and subsequently taught there himself while continuing study for a bachelor of arts in mathematics at the University of Michigan. In 1902, he moved to New York and began teaching mathematics at the Horace Mann School, which was affiliated with Teachers College. After receiving the master's degree and completing further study at the University of Göttingen, Upton joined the Teachers College faculty as an instructor in the mathematics department. He remained at the college for the rest of his career.

The Doctoral Program

Within two years of his arrival at Teachers College, Smith had a master's degree program with a thesis component firmly in place. Prototype doctoral programs in education had been introduced a few years earlier by other departments at Teachers College. With the full support of Dean Russell, Smith formulated a plan for a Ph.D. program in the history and teaching of mathematics. As with sister programs, components of Smith's program were categorized into three general areas: course work, examinations, and the dissertation.

Course Work

Requirements included a minimum two-year residency and completion of sixteen graduate courses. The courses were to be distributed among general education, higher mathematics, and the history and teaching of mathematics. The general education category included courses in the history, philosophy, and psychology of education, in keeping with Russell's (1900a) view of the importance of this broader "professional" knowledge.

Some mathematics courses were offered at Teachers College, but courses in higher mathematics usually were taken at Columbia University. Smith urged all his graduate students to include as much mathematics, both pure and applied, as possible. He recommended pure mathematics courses such as algebra and the theory of equations, projective geometry, theory of functions, and advanced calculus. In applied mathematics, Smith suggested analytic mechanics and astronomy.

Preparation in the history and teaching of mathematics required completion of at least six courses. The theory and practice course used Smith's (1900) recently published book as a text. The course included comparisons of mathematics curricula and textbooks in the leading countries. Two courses in the history of mathematics, introductory and advanced, dealt with secondary school and college topics. Both courses emphasized use of original sources and required a reading knowledge of French and German. The history courses were cross-listed under the Columbia mathematics department to encourage wider enrollment. A practicum course was offered on a continuing basis. Central to the practicum was discussion of issues in mathematics education, for example, methods of beginning the differential calculus. The introductory history course was a prerequisite for enrollment in the practicum. Masters students enrolled in one semester of the practicum, while doctoral students were required to enroll for at least two semesters. The practicum resulted in a report that usually served as the master's essay. Doctoral students completed the required course work by enrolling in a seminar to prepare the dissertation (Teachers College, 1902-1903; Teachers College Department of Mathematics, 1906-1907, 1911-1912).

Examinations

Doctoral students also had to pass examinations on general education topics, on mathematics and education, and on two languages, French and German. The examinations involved both written and oral segments.

Dissertation

The final component of the Ph.D. program was the preparation of a dissertation that showed "evidence of original investigation and research" (Teachers College, 1902, p. 5). An oral defense of the manuscript before a panel of faculty was the final step for the degree. Early dissertations were published in a Teachers College series entitled *Contributions to Education* (for a discussion of research in mathematics education during this period see Kilpatrick, 1992).

First Graduates

In 1906, Teachers College awarded the first Ph.D. degrees in mathematics education to Lambert Lincoln Jackson and Alva Walker Stamper. Jackson, who was 36 years old when he received the degree, had been Smith's protégé since his student days at Cortland Normal School. When Smith moved from Cortland to Ypsilanti, Michigan, he invited Jackson to join him as his assistant at the normal school. Jackson combined his work for Smith with further study at the University of Michigan. He completed a B.A. and M.A. in mathematics and began study for the Ph.D. When Smith took the principalship of Brockport Normal, he offered Jackson a position as head of the mathematics department there. Jackson accepted the offer. Smith left Brockport for New York in 1901; Jackson remained at Brockport and kept in contact with his mentor. Subsequently, he began collaborating with J. W. A. Young to produce a series of arithmetic texts (Young and Jackson, 1904).

Beginning in 1903, Jackson enrolled at Teachers College to continue study for a doctoral degree. While working on the dissertation, he traveled from Brockport to New York to do research, sometimes staying in Smith's home. Jackson chose as a dissertation topic the history of arithmetic (Jackson, 1906). At least two factors influenced this choice: his collaboration with Young on the arithmetic series and his access to primary sources. Jackson could use the extraordinary collection of sixteenth-century printed arithmetics being assembled at the time by George A. Plimpton. Smith served as Plimpton's advisor in this pursuit and arranged for student access to the collection (see Donoghue, 1998). After receiving the Ph.D., Jackson took a position in publishing with Appleton and Company, which had published the Young-Jackson arithmetics. Subsequently, he served as assistant commissioner of education for the State of New Jersey.

Unlike Jackson, Stamper did not know Smith prior to enrolling at Teachers College in 1904. He was 33 years old and had journeyed to New York from California. Stamper began his career as the principal of an elementary school, a post he held for two years. In 1895, he received a bachelor of science degree from the University of California in Berkeley. His major was mathematics, but he included enough work in pedagogy to obtain a certificate to teach. For three years, Stamper taught mathematics at Berkeley High School while pursuing a doctorate at the university.

Before completing the doctorate, Stamper accepted a position as head of the mathematics department at Chico State Normal School. His experiences there changed his intentions and his goals. Stamper taught a course for teachers at the university during the 1902 summer session. He used Smith's book as a guide, even borrowing its title for the course name, Teaching of Elementary Mathematics. As a result, Stamper developed an interest in the history of mathematics and recognized the necessity of further preparation for his work with teachers. In a letter of inquiry to Smith dated 14 July 1903, Stamper explained his situation. He emphasized that although he wished "to take some pure mathematics, as well as work that will strengthen me in the general subject of Education," he did not wish to major simply in education: "To sum up, I want my major to be in *Mathematical* Education" (Smith Papers).

Stamper (1909) entered Teachers College as a special graduate scholar and earned an M.A. and a Ph.D. His dissertation topic was *A History of the Teaching of Elementary Geometry with Reference to Present Day Problems*. Stamper used original sources in Smith's own library as well as relevant works in Plimpton's collection. He also traveled to Europe in 1905 to conduct further research. After completing the doctorate, Stamper returned to Chico, where he remained for a number of years.

Jackson and Stamper had similar professional backgrounds. Each had begun doctoral study in mathematics but revised his intentions as a result of his experiences teaching at a normal school. Each chose a historical topic for research, no doubt influenced by Smith's expertise and by the unusual availability of rare primary sources. However, their individual interests led them toward these topics as well. In each dissertation the author situates his study within the existing literature, establishes a need for the study, and explains how his work is an original and unique contribution to scholarship.

Only one other American dissertation in the area of mathematics education was completed prior to 1912. Cliff Winfield Stone (1908) conducted a statistical study of test performance by 6,000 sixth-grade students in 26 school systems in the East and Middle West. He titled his work *Arithmetical Abilities and Some Factors Determining Them.* Stone examined abilities in performing the four basic operations of addition, subtraction, multiplication, and division. The factors in question were expenditure of time and the nature and arrangement of topics. Edward L. Thorndike guided the statistical aspects of the study, David Eugene Smith supervised test design, and George Strayer, a specialist in elementary school education, served as overall advisor. By 1916, Stone had become the director of teaching at Iowa State Teachers College in Cedar Falls.

UNIVERSITY OF CHICAGO

At the turn of the twentieth century, the University of Chicago was the leading center of mathematical research and study in the United States, due in large part to the leadership and determination of E. H. Moore. The graduate programs in mathematics education were developed with Moore's guidance and cooperation by J. W. A. Young in the mathematics department and George Myers in the school of education.

Characteristics of the Chicago Program

In his 1902 address as retiring president of AMS, Moore (1903/1926) expressed his regret that a chasm had developed between pure and applied mathematics. As a partial remedy Moore recommended major reform of school mathematics along the lines of the Perry movement. John Perry (1901), a British professor of mechanics and mathematics, had proposed that practical mathematics be taught using the scientific method. Moore proposed that algebra and geometry be "fused" with the physics curriculum into a coherent sequence taught using laboratory methods. Moore believed that mathematics should become an experimental science.

The fusion of mathematics and physics and the laboratory method of teaching became two hallmarks of the university's program to train mathematics teachers. To achieve his goal, Moore enlisted two colleagues, Young and Myers. Moore had recruited Young, a Ph.D. graduate of Clark University, to join his new department in 1892. Six years later, Young was promoted from instructor of mathematics to assistant professor of mathematical pedagogy, the first title of its kind in the United States. Myers was on the faculty of the university's newly formed school of education that resulted from consolidation with the Chicago Institute.

Moore, Young, and Myers cooperated in preparing mathematics teachers for the secondary schools and colleges. Moore himself designed and taught a course for teachers on the use of graphical methods in algebra, particularly the use of graph paper. Young regularly offered courses such as critical review of secondary mathematics for teachers. The extensive selection of courses offered by the mathematics department at the time was exceptional in the United States (University of Chicago, 1904-1905, 1906-1907).

Courses taught by Myers in the school of education through 1912 were designed to prepare teachers or supervisors of mathematics at the elementary, secondary, or normal school levels. Occasionally, during a summer quarter, Young taught more advanced courses that enrolled practicing teachers. The courses focused upon content, applications, or pedagogy. Content courses offered through the school of education included analytics, history of mathematics, and a course in secondary and collegiate mathematics. Applications courses dealt with astronomy, surveying, mathematical geography, mechanics, and the use of calculus to solve physical problems. The pedagogy courses included offerings in the theory and practice of teaching mathematics (University of Chicago, 1902-1903, 1904-1905, 1906-1907).

Master's Degrees

In 1910, Myers offered a new pedagogy course, teaching of secondary mathematics in Europe (University of Chicago, 1909-1910). For the first time in the school of education, a mathematical pedagogy course was designated as intended primarily for graduate students. A second course for graduate students added later that year, problems in mathematical education, was a research course that examined topics such as critical and historical studies of the growth of mathematics, justification for secondary programs, comparisons with programs in other countries, "modernized" curriculum and methodology, critiques of significant texts, and methods for studying the "results" of mathematical teaching (University of Chicago, 1910-1911, 1911-1912).

Two students, Walter Clifton Erwin and Christine Bednar, produced master's theses in mathematics education (Erwin, 1910; Bednar, 1910). Erwin's thesis for a master of arts degree, "Study of a Preparatory School Class in Algebra," was based upon his experience as a mathematics teacher in the preparatory department at the University of Oklahoma. Universities and colleges established preparatory departments in the latter half of the nineteenth century to remedy deficiencies in the academic preparation of students who had attended rural or village schools. Although the intention was to prepare students for college-level work, not all preparatory graduates went on to enroll in colleges. First-year preparatory students usually were older than the typical first-year high school student; this was the case in Erwin's class. He studied nine males and four females to determine individual differences in their algebra performance. Erwin reported his findings as micro case studies. Each student's work was profiled, and reasons for the level of performance were discussed. In the process, Erwin offered a critique of the text he used for the course, Milne's *High School Algebra.*

Christine Bednar received a master of science degree with a joint major in mathematics and education. Her thesis, "Educational Grounds for Unified Mathematics," examined the justification for using a "modernized" curriculum in the schools. The topic reflects the influence of Moore, Young, and Myers.

From 1900 to 1920, fifteen master's theses in mathematics education were completed at the University of Chicago. Five of the theses examined aspects of elementary arithmetic. The remaining ten theses concerned issues for secondary schools. Four of the ten theses on secondary school issues dealt with teaching and/or learning, two with curriculum, and two with text analysis (University of Chicago, 1900-1920).

Introduction of a Doctoral Program

Concurrent with the work of Erwin and Bednar, the mathematics department at Chicago expanded the areas for doctoral research beyond pure and applied topics to include study of the history, philosophy, or pedagogy of mathematics (University of Chicago, 1909-10). In 1911, Theodore Lindquist took advantage of that expansion. An Illinois native, Lindquist had received an A. B. in 1897 from Lombard College in Galesburg, near his home. After earning an M. S. in 1899 at Northwestern University, he taught mathematics at the secondary school level for a brief period.

While at Chicago, Lindquist worked and studied in the departments of mathematics and physics. Over six academic quarters, he took courses from physicists A. A. Michelson and Robert A. Millikan, both eventual Nobel laureates, and from renowned mathematicians G. A. Bliss, Oskar Bolza, Leonard Dickson, and Heinrich Maschke. Although Lindquist undertook a program weighted heavily toward the scientific-technical side, for his thesis he chose to analyze the state of engineering education in the United States. The thesis, "Mathematics for Freshman Students of Engineering", included a history of engineering preparation, an examination of the mathematics curriculum at engineering colleges, and recommendations for curricular and pedagogic improvements at the secondary and college levels. J. W. A. Young directed the thesis. Although Lindquist's thesis dealt with topics of an educational nature, he received a Ph.D. in mathematics.

By 1912, Chicago's mathematics department recommended that all prospective college and university teachers take courses in the history of mathematics and in the principles and practice of education. Graduate students were encouraged to observe college classes in mathematics and, if possible, to procure positions as teaching assistants (University of Chicago, 1911-12).

In 1915, Walter Scott Monroe completed Chicago's first doctorate in mathematics education. For the dissertation, he chose a topic that combined history and pedagogy, the influence of Warren Colburn and the Pestalozzi movement upon the development of school arithmetic. Myers served as dissertation advisor.

Prior to enrolling at Chicago, Monroe had earned three degrees at the University of Missouri, an A.B. (1905), an S.B. (1907), and an A.M. (1911). W. W. Charters was Monroe's mentor at Missouri, and it was Charters who had suggested to Monroe the topic of Colburn's influence on arithmetic in America. After receiving the Ph.D. from Chicago, Monroe went on to a prominent career in educational research. He first took a position as director of the Bureau of Cooperative Research at Indiana University. Within a few years, he decided to join Charters, who had moved from Missouri to the University of Illinois. At Illinois, Monroe held a faculty appointment in the college of education and directed numerous studies through the university's Bureau of Educational Research.

Chicago did not award another Ph.D. in mathematics education until 1922. The recipient, Karl John Holzinger (1922), completed a dissertation entitled *The Indexing of a Mental Characteristic.* By that time, the young mathematics education community already had engaged in its first major battle, a defense against external attacks by factions of the progressive education movement (see Stanic, 1986). The more extreme reformers in the movement demanded that mathematics educators justify a place in the secondary school curriculum for any topic beyond practical grammar school arithmetic. Their paramount concern was the social benefit of studying mathematics. Thus imperiled, mathematics educators realized that differences among themselves concerning curriculum or methodology were insignificant in comparison with the values they shared: a regard for the power and beauty of mathematics as an autonomous discipline, an appreciation of its myriad applications, and a dedication to the improvement of mathematics teaching at every level.

FROM THE SEEDS OF EARLY DOCTORAL PROGRAMS IN MATHEMATICS EDUCATION

The field of mathematics education, nurtured by a handful of institutions in the 1890s and early 1900s, was cultivated in a broader spectrum of universities, colleges and normal schools by master's and doctoral graduates of Teachers College and the University of Chicago. Eventually, universities across the United States established their own graduate programs in mathematics education. Although today's programs may differ from one another in particular ways, all share the heritage bequeathed by those who pioneered the field a century ago.

Eileen F. Donoghue
College of Staten Island, City University of New York
3S-208
2800 Victory Blvd.
Staten Island, NY 10314
donoghue@postbox.csi.cuny.edu

CBMS Issues in Mathematics Education
Volume 9, 2001

DOCTORAL PROGRAMS IN MATHEMATICS EDUCATION IN THE UNITED STATES: A STATUS REPORT

Robert E. Reys, University of Missouri
Bob Glasgow, Southwest Baptist University
Gay A. Ragan, Southwest Missouri State University
Kenneth W. Simms, University of Missouri

INTRODUCTION

Mathematics education has long struggled to define itself as an academic discipline and research area. Details of this saga have been traced and documented (Kilpatrick, 1992) but the lack of self-identify continues to both challenge and frustrate mathematics educators (Sierpinska & Kilpatrick, 1998; Silver & Kilpatrick, 1994). It is ironic that while the field of mathematics education lacks a clearly articulated identity, the roles of and demand for mathematics educators continue to expand (Reys, 2000).

The role of mathematics educators has evolved and continues to evolve over time. Historically, that role tended to be as faculty in (1) mathematics departments whose primary function was to teach mathematics and/or to prepare mathematics teachers; and (2) colleges of education where there was the expectation of research related to mathematics education. A newly emerging job-placement for doctoral graduates in mathematics education is within mathematics departments actively seeking mathematics educators not only to teach mathematics and mathematics education classes, but also to engage in and lead research on the teaching and learning of undergraduate mathematics. This trend confirms a growing acceptance and expectation of scholarship based on research in mathematics education for faculty in mathematics departments. In addition to careers in higher education, there are increasing opportunities for people with doctorates in mathematics education in large school districts (including jobs as both classroom teachers and district coordinators) as well as in leadership roles in city, regional, and state departments of education.

These diverse and growing employment opportunities for mathematics educators have created a demand that exceeds the current supply of people with doctorates in mathematics education. The shortage has been exacerbated by a multitude of factors, including the aging of college/university faculty in mathematics and mathematics education; new areas of specialty in mathematics education due to the rapid and dramatic changes in technology to support the learning and teaching of mathematics; a decrease in the number of students studying advanced mathematics in higher education; a shortage of certified mathematics teachers in middle, junior, and senior high schools; and increasing mathematics requirements at the secondary and post-secondary levels.

A clear characterization of the discipline of mathematics education would seem to be a prerequisite for advanced study leading to a doctorate in mathematics education; yet, despite the lack of consensus on what constitutes mathematics education, doctoral programs in mathematics education have evolved throughout the twentieth century (Donoghue, this volume). These programs have taken many different forms, as the focus of preparation varies greatly from one institution to another. While one program may consist of a large component of mathematical content together with selected graduate courses in education, another program may offer a more balanced set of courses in mathematics and mathematics education. Some programs use a mentoring model involving graduate students in research projects, thereby providing research practicums or internships, while others provide research preparation via course work only. One program may graduate a cadre of students annually with doctorates in mathematics education, while other programs produce one graduate every few years. Each program is autonomous and seemingly oblivious to other mathematics education programs around the country. All of these factors contribute to very diverse standards of preparation for people holding doctorates in mathematics education. The challenges of characterizing doctoral programs in mathematics education are closely related to, and perhaps only surpassed by, the difficulties encountered in defining mathematics education as a discipline.

While data reporting trends in the quantity of doctorates in the mathematical sciences have been reported (National Research Council, 1998), no parallel data are available for doctorates in mathematics education. In an effort to gather information about current doctoral programs in mathematics education a survey was conducted. This paper outlines the methods used to collect information regarding doctoral programs in mathematics education and reports the results.

IDENTIFICATION OF PROGRAMS

Our goals in the survey were to learn which institutions offer doctoral programs with a major area in mathematics education, and to gather information about their faculties, students, and programs. Thus, related questions asked about each institution included: How many mathematics education faculty contribute to the program? How many doctoral students are in the program? How many doctoral students graduate annually? What program of study does the institution require? Answers to these questions would provide a snapshot of doctoral programs in mathematics education.

It seems reasonable that doctoral programs of the sort we wanted to learn about would produce students who did their dissertation research in mathematics education. Therefore, if these dissertations could be identified, it would be easy to link them to the specific universities awarding the degrees and determine which institutions were producing doctorates in mathematics education. That task turned out to be much easier said than done.

Our first approach was to examine the annual listings of doctoral dissertations published in each July issue of the *Journal for Research in Mathematics Education (JRME)*. The keywords used in the search were "Education: Mathematics." The July 1994 *JRME* listed dissertations for 1993. However, this was the last annual listing of dissertations in *JRME*, so a separate search was conducted using the Dissertations Abstract (DISS) database for the years 1994 to 1997. This process yielded dissertations published by institutions over the 18-year period from 1980 to 1997. An examination of the listings revealed that although these dissertations were related to mathematics education, the studies reported were not all carried out by people specializing in mathematics education. In fact, people other than mathematics educators did most of the dissertations listed in *JRME*. A wide range of disciplines was represented including educational psychology, sociology, technology, elementary education, administration, counseling, and special education. It appears that mathematics teaching and learning provides the context for doctoral research in these other areas. While this annual listing of dissertations had seemed like a good source to identify doctorates in mathematics education, we felt another approach was needed.

Next we examined two documents prepared by the National Research Council (NRC). The *Summary Report 1996: Doctoral Recipients from United States Universities* and the *Summary Report 1997: Doctoral Recipients from United States Universities* provide data on conferred research doctorates from almost 400 colleges and universities in the United States (NRC, 1998; 1999). Data are collected through surveys and distributed via the graduate deans of each institution, to students who are completing requirements for their doctorate. These reports summarize research and applied-research doctorates in all fields, including Ph.D., D.Sc., and Ed.D., with the Ph.D. label being generically used to refer to any of these degrees. While not all parts of the survey are completed by every graduate, all doctoral recipients provided their gender, Ph.D. major field, and year (NRC, 1998, p.2).

One item on the NRC survey asks graduates to identify the field of specialization for their Ph.D. field from a given list of codes. Graduates who identified "Mathematics Education" as their major area (code 874) were considered to have earned a doctorate in mathematics education. It can be argued that self-reporting provides reliable and valid data (who better to identify her/his major field than the person receiving the doctorate). However, the organizational structure of an institution may lead some mathematics educators to code their area of emphasis in other ways, such as "Curriculum and Instruction" or "Elementary Education" and therefore to be lost from this survey. The resulting list of 'mathematics education doctorates' may be conservative; yet, it provides an identifiable list of people who classified mathematics education as the major field for their doctorate.

The *Summary Reports* disclosed that 100 doctorates were earned in mathematics education (by 35 males and 65 females) from July 1, 1995 to June 30, 1996 and 88 (by 37 males and 51 females) from July 1, 1996 to June 30, 1997. The institutions conferring these degrees were not identified in the *Summary Reports*. However, we requested from the NRC the names of institutions from which doctorates in mathematics education were received during the period from 1980 to 1997. These data are summarized and provide the basis for discussion in this paper. (It should be noted that after the survey was completed, but prior to publication of this paper, the 1998 NRC data were received and those results are reflected only in table 3).

NATIONAL RESEARCH COUNCIL SUMMARY REPORTS

The *Summary Reports* provide different perspectives of doctoral recipients, including gender, age, ethnicity and specialty area. Our examination of these data included mathematics and education as separate categories but focused primarily on mathematics education.

DOCTORATES IN MATHEMATICS

How does the number of doctorates in mathematics education compare to the number of doctorates in mathematics? Table 1 reports the breakdown of doctorates in different areas of mathematics as reported in the NRC *Summary Reports*. These data suggest, as a comparison, that about as many doctorates are earned in the subfield of Analysis each year as in Mathematics Education (see table 2). These data also provide interesting benchmarks for discussion that will resurface when information about the job market is addressed.

Table 1. Number of Doctorate Recipients in Mathematics by Subfield in 1995–1996 and 1996–1997

	1995–1996			1996–1997		
Subfield	Total	Men	Women	Total	Men	Women
Applied Mathematics	230	178	52	241	186	54
Algebra	78	60	18	79	58	21
Analysis and Functional Analysis	100	85	15	103	90	13
Geometry	72	58	14	70	57	13
Logic	16	15	1	23	18	5
Number Theory	42	35	7	46	35	11
Mathematical Statistics	178	131	47	182	134	46
Topology	55	50	5	62	48	14
Computing Theory and Practice	18	16	2	14	13	1
Operations Research	21	17	4	20	15	5
Mathematics, General	233	188	45	143	98	41
Mathematics, Other	79	58	21	129	93	36
Mathematics—All Areas	**1122**	**891**	**231**	**1112**	**845**	**260**

Note. The data in Table 1 are from *1996 Summary Reports: Doctoral Recipients from United States Universities* (p. 66) and from *1997 Summary Reports: Doctoral Recipients from United States Universities* (p. 86), by NRC, 1998 and 1999, Washington D.C.: National Academy Press.

DOCTORATES IN EDUCATION

Table 2 shows the number of doctorates in the education categories of the *Summary Report 1996* and the *Summary Report 1997*. This table provides a panoramic view of the range of specialty areas at the doctoral level in education and illustrates that the number of doctorates awarded in different specialty areas varies greatly. While mathematics education is reported as a separate specialty field, it is likely that other people with considerable interest and expertise in mathematics education are counted in some other fields such as Curriculum and Instruction, Elementary Education, or Secondary Education. The data in table 2 show that the majority of recipients of doctorates in education, including mathematics education, are female. While 116 of the 188 (62%) of the doctorates awarded in 1996 and 1997 in mathematics education went to females, table 1 reports that only 491 of the 2,234 (22%) of the doctorates awarded in mathematics over the same time period went to females.

Table 2. Number of Doctorates in Education and Teaching Fields for 1995–1996 and 1996–1997

EDUCATION	1995–1996			1996–1997		
	Total	Men	Women	Total	Men	Women
Curriculum and Instruction	896	266	630	904	278	623
Educational Administration & Supervision	1170	535	635	1020	427	585
Educational Leadership	989	428	561	1036	423	607
Educational/Instruction Media Design	107	47	60	92	38	53
Educational Statistics/Research Methods	76	34	42	58	18	40
Educational Assessment, Tests & Measure	32	19	13	29	12	17
Educational Psychology	309	90	219	356	114	241
School Psychology	114	33	81	115	35	80
Social/Philosophical Foundations	125	44	81	135	51	82
Special Education	278	64	214	263	47	216
Counseling Education/Counsel/Guidance	277	93	184	203	72	131
Higher Education/Evaluation & Research	481	205	276	509	215	292
Pre-elementary/Early Childhood	81	13	68	42	6	36
Elementary Education	46	6	40	54	9	45
Secondary Education	34	10	24	25	11	14
Adult and Continuing Education	210	86	124	161	64	97
TEACHING FIELDS TOTAL	863	361	502	894	346	545
Agricultural Education	32	22	10	38	26	11
Art Education	41	15	26	29	8	20
Business Education	20	9	11	22	11	11
English Education	57	15	42	60	16	44
Foreign Languages Education	44	15	29	45	12	33
Health Education	90	29	61	59	16	42
Home Economics Education	13	0	13	13	3	10
Technical/Industrial Arts Education	11	7	4	19	10	9
Mathematics Education	**100**	**35**	**65**	**88**	**37**	**51**
Music Education	91	46	45	98	46	52
Nursing Education	23	0	23	22	1	21
Physical Education & Coaching	101	60	41	108	52	56
Reading Education	66	13	53	68	10	58
Science Education	96	50	46	73	33	40
Social Science Education	12	5	7	27	12	15
Technical Education	24	20	4	32	15	17
Trade & Industrial Education	12	7	5	16	8	8
Teacher Ed./Spec. Acad. & Voc.	30	13	17	77	30	47
Education, General	353	141	212	317	116	181
Education, Other	331	118	213	284	85	194
EDUCATION GRAND TOTAL	6772	2593	4179	6497	2367	4079

Note: The data in Table 2 are from *1996 Summary Reports: Doctoral Recipients from United States Universities* (p. 67) and from *1997 Summary Reports: Doctoral Recipients from United States Universities* (p. 87), by NRC, 1998 and 1999, Washington D.C.: National Academy Press.

DOCTORATES IN MATHEMATICS EDUCATION

Table 3 focuses on the subset of doctorates in mathematics education. It reports every institution (a total of 126) that awarded at least one doctorate to someone who identified mathematics education as her/his major field from 1980 to 1998 as given by the NRC *Summary Reports*. Table 3 also reports the number of doctorates awarded annually by each of these institutions. These numbers vary across years, but the total does not exceed 115 in any given year.

Table 3. Number of Mathematics Education Doctoral Degrees by Year (1980–1998*)

Institutions	'80	'81	'82	'83	'84	'85	'86	'87	'88	'89	'90	'91	'92	'93	'94	'95	'96	'97	'98	Totals
Teachers College-Columbia U/NY	8	3	3	4	6	5	6	5	5	7	9	7	5	5	5	8	8	4	9	112
University of Georgia	3	2	2	6	5	3	7	2	3	3	2	8	2	6	5	10	10	5	5	89
University of Texas-Austin	4	3		9		2	4	4	5	2	3	9	4	6	2	3	3	5	11	79
Ohio State University	1	5	2		3	2	4	3	2	1	1	4	3	5	5	5	5	3	5	59
Georgia State University	2	1	3	2	2	1	2	4	3	5	2	1	2	3	5	5	2	1	6	52
New York University	6	4	2	5	5	3	2	4	2		5	1	1	1		1	2		2	46
Florida State University	2	5	2		2	2	4			2	1	2	4	2	3	1	1	2	3	38
University of Maryland	2	2		2		4	4	3	4		2	2		3	1		2	2	4	37
Temple University/PA	2	2	4	1	1	1	3	2	3	2	3	2			1		3	2	1	33
Rutgers University/NJ	1	4	2	3	2			2	1	1	5	1	1		1	4	2	1	1	32
Indiana University-Bloomington	3	3	2	2	1	2		3	1	1		2	1	3	1		2	2	2	31
University of Iowa	3	3		1		2	1	3	2	3		3	2	1	3	1	2		1	31
State U of New York-Buffalo	1	2	1		2	1	1	1		3	2	2	3	2	1	2	2	1	2	29
University of Wisconsin-Madison	5		1		1	2	2	1			1		1	1	4		4	4	2	29
Boston University/MA			2	3	4			3	3	3	2	1	1				3	2		27
American University/DC			2			1	1	1	1	4	2	1	1	2	1	3	1	2	2	25
University of Pittsburgh/PA	1			1	3			2	3	4	1	2			4	2			2	25
Columbia University/NY	4		1	2		4	2	2			2			1		1		2	2	23
North Carolina State U-Raleigh		1			2		1	1	1	2			2	1	1	2	1	5	3	23
University of Oklahoma	1	2	1				1	1	1		1		2	1	2	4	2	1	3	23
University of California-Berkeley	1	1	1	2	2		3		3		3		1			1	3			21
University of Minnesota-Twin City	3		1			1		5						2	1		3	2		18
University of Missouri-Columbia	1	2			2	5	1	1			2	1		1	1				1	18
U of Massachusetts-Amherst	2		1			1	2	1	1	2		1	3	2		1				17
Michigan State University		1	1	2	2	2	1		1	1				1	1		1	2		16
Oregon State University		1	1		1		1	1		1		3		1		1	2	1	1	15
U of Illinois-Urbana-Champaign		1			2		2	2	1	1	2	1			2		1			15
University of Tennessee-Knoxville				1	1	2	1	1			2	1	1		3	1		1		15
Peabody Col of Vanderbilt/TN			2		1		1			1	3	1		1	1	1	2			14
University of South Florida		1	2			1		1		1		2			1	1	1	2	1	14
Syracuse University/NY					1		2		1	1	1				2	3		2		13
University of South Carolina				1				1			1	4	1	2			2	1		13
Illinois State University-Normal					1			1						2	3	3	2			12
Purdue University/IN			1	1	2		2		1	1		1		2			1			12
Auburn University/AL	2	1				1	1	1			1			1	1	1		1		11
Cornell University/NY					1				1		1	1	1	2	2		1	1		11
University of Houston/TX	1		2	2			1			1		1	2			1				11

(continued)

Table 3. (continued)

Institutions	'80	'81	'82	'83	'84	'85	'86	'87	'88	'89	'90	'91	'92	'93	'94	'95	'96	'97	'98	Totals
University of Northern Colorado	2		1											1		2	3	1	1	11
University of Virginia	1	1			1		2					2	1		1			1	1	11
University of Michigan-Ann Arbor			1		1							1			1	3		1	2	10
Pennsylvania State University		1	1		2	1			1			1					1		1	9
Stanford University/CA	1	2	1	1										1		2		1		9
University of Florida	1			2		1		1							2	2				9
University of Mass-Lowell														2	1	1	2	1	2	9
Northwestern University/IL	1	1		2	1	1							1			1				8
Ohio University			2					1	1				1				2	1		8
Oklahoma State University		1	1		1	1	1		1						2					8
Southern Illinois University							1	1			1	1	1	1	1	1				8
University of Delaware					1											2	1	2	2	8
University of Alabama										2					2	1		1	1	7
University of New Hampshire	1	1			1		2						1				1			7
Washington State University														1	1		2		3	7
Texas A&M University							1		1						3		1			6
University of Chicago/IL													3					1	2	6
University of Oregon	3	1	1												1					6
Claremont Graduate School/CA										1			1				1	1	1	5
University of Denver/CO		1				1		3												5
West Virginia University			1		2							1				1				5
Kansas State University			1				1		1	1										4
Montana State University										1					2		1			4
U of Southern Mississippi	1	1								1							1			4
University of Colorado		1			2						1									4
University of Toledo/OH										1					1		1	1		4
University of Washington			1		1										1		1			4
Vanderbilt University/TN										1								1	2	4
Wayne State University/MI				1	1													1	1	4
Kent State University/OH													1		2				1	4
Florida Atlantic University					1				1	1										3
Harvard University/MA							1			1						1				3
State U of New York at Albany	1									1									1	3
U of North Carolina-Greensboro															1	2				3
University of Arizona											1								2	3
University of Cincinnati/OH				1								1			1					3
University of Connecticut		1	1		1															3
University of Mississippi													2		1					3
University of Rochester/NY	1			1	1															3
Western Michigan University															1			2		3
Arizona State University	1					1														2
Bowling Green State University/OH								2												2
Clark University/MA						1									1					2
CUNY Grad. School & Univ. Center																			2	2
Emory University/GA						1						1								2
George Mason University/VA											1						1			2

(continued)

Table 3. (continued)

Institutions	'80	'81	'82	'83	'84	'85	'86	'87	'88	'89	'90	'91	'92	'93	'94	'95	'96	'97	'98	Totals
Lehigh University/PA								1	1											2
Louisiana State U & A&M College													1		1					2
Northern Illinois University													1					1		2
U of Missouri-Saint Louis									1									1		2
U of North Carolina-Chapel Hill																	1	1		2
University of Illinois-Chicago													1			1				2
University of Kansas																	1	1		2
University of Nebraska-Lincoln						1										1				2
University of New Mexico										1						1				2
University of North Dakota																2				2
University of Southern California					1											1				2
Utah State University				1												1				2
Walden University/MN																	1	1		2
Ball State University																		1		1
Baylor University/TX													1							1
Boston College																			1	1
Brandeis University																			1	1
Brigham Young University/UT			1																	1
Duke University/NC								1												1
Fairleigh Dickinson University/NJ		1																		1
Florida Institute of Technology																1				1
Memphis, University of/TN															1					1
Miami University - Oxford																			1	1
Mississippi State University								1												1
Northwestern State University/LA										1										1
Nova Southeastern University/FL	1																			1
South Carolina State University														1						1
State U of New York-Binghamton																1				1
Tennessee State University																			1	1
U of Missouri-Kansas City																1				1
University of Akron/OH						1														1
University of California-Davis																		1	1	1
University of Kentucky																1				1
University of Miami/FL												1								1
University of Montana																1				1
University of Notre Dame																			1	1
University of Sarasota																			1	1
University of South Dakota															1					1
University of Wisconsin-Milwaukee								1												1
University of Wyoming		1																		1
Virginia Polytech Inst & State U													1							1
Washington University																			1	1
University of Central Florida																				0
TOTALS	74	62	50	62	64	65	72	74	56	69	65	73	62	69	74	92	100	88	115	1386

Note. Data are from National Research Council upon special request. Completed surveys were received from institutions in **bold**. *Italicized* institutions responded that they do not currently offer a doctoral degree in mathematics education.
*1998 data was not received until after the survey was completed and thus was not used to group institutions (see table 4).

Table 4 compares the number of graduates as reported by the NRC to the number of dissertations related to mathematics education listed in the *JRME* or in the Dissertation Abstracts database (DISS) from 1980 to 1997. While the number of dissertations reported in *JRME* or DISS exceeds the number of reported graduates by the NRC for every institution, their Pearson's correlation coefficient is 0.836. As noted in table 4, 1,271 mathematics education doctorates were awarded from 1980 to 1997 from these 117 institutions. During the same period, 4,157 dissertations related to mathematics education were documented from the same 117 institutions. The one to three ratio of graduates with doctorates in mathematics education versus dissertations related to mathematics education (NRC to *JRME*) is a vivid reminder of the amount of doctoral research in mathematics education done by people in fields other than mathematics education.

An examination of table 3 shows that over the 18-year time period, the vast majority of institutions were producing less than one doctorate in mathematics education each year. Only the five largest programs (Teachers College, University of Georgia, University of Texas, Ohio State University, and Georgia State University) averaged more than three graduates annually over the last decade. Pooling the data from the 15 largest programs (the previous five plus New York University, Florida State University, University of Maryland, Temple University, Rutgers University, University of Iowa, Indiana University, State University of New York at Buffalo, University of Wisconsin-Madison, and Boston University) show that these institutions (hereafter referred to as Group 1 and shown in table 5) produced about half of all doctorates awarded during this period: 724 out of 1,386.

Table 4. Number of Mathematics Education Doctoral Degrees by Year (1980–1997) and Dissertation Totals

Institutions	NRC Totals	JRME/DISS Totals	Institutions	NRC Totals	JRME/DISS Totals
Teachers College-Columbia U/NY	103	133	U of Massachusetts-Amherst	17	54
University of Georgia	84	140	University of Missouri-Columbia	17	57
University of Texas-Austin	68	121	Michigan State University	16	60
Ohio State University	54	106	University of Minnesota-Twin City	16	43
Georgia State University	46	76	U of Illinois-Urbana-Champaign	15	62
New York University	44	64	Peabody Col of Vanderbilt/TN	14	40
Florida State University	35	82	University of Tennessee-Knoxville	14	58
University of Maryland	33	67	Oregon State University	14	32
Temple University/PA	32	104	University of South Florida	13	43
Rutgers University/NJ	31	79	Purdue University/IN	12	35
University of Iowa	30	84	University of South Carolina	12	64
Indiana University-Bloomington	29	69	Syracuse University/NY	11	29
State U of New York-Buffalo	27	55	University of Houston/TX	11	66
University of Wisconsin-Madison	27	104	Auburn University/AL	10	32
Boston University/MA	25	53	Cornell University/NY	10	24
University of Pittsburgh/PA	23	72	University of Northern Colorado	10	44
American University/DC	23	31	University of Virginia	10	32
Columbia University/NY	21	63	Illinois State University-Normal	10	33
University of California-Berkeley	21	43	University of Florida	9	57
University of Oklahoma	20	31	Stanford University/CA	9	57
North Carolina State U-Raleigh	20	31	Northwestern University/IL	8	23

(continued)

Table 4. (continued)

Institutions	NRC Totals	JRME/DISS Totals	Institutions	NRC Totals	JRME/DISS Totals
Oklahoma State University	8	29	State U of New York at Albany	2	25
Southern Illinois University	8	27	University of Illinois-Chicago	2	16
Pennsylvania State University	8	54	University of Nebraska-Lincoln	2	42
University of Michigan-Ann Arbor	8	48	University of New Mexico	2	8
University of New Hampshire	7	10	University of North Dakota	2	7
University of Mass-Lowell	7	?	University of Southern California	2	80
Ohio University	7	19	Utah State University	2	11
Texas A&M University	6	57	Walden University/MN	2	10
University of Oregon	6	53	Northern Illinois University	2	35
University of Alabama	6	31	U of Missouri-Saint Louis	2	7
University of Delaware	6	19	U of North Carolina-Chapel Hill	2	34
University of Denver/CO	5	21	University of Kansas	2	29
West Virginia University	5	29	Vanderbilt University/TN	2	12
Kansas State University	4	33	Baylor University/TX	1	14
Montana State University	4	21	Brigham Young University/UT	1	32
Univ of Southern Mississippi	4	45	Duke University/NC	1	5
University of Colorado	4	29	Fairleigh Dickinson University/NJ	1	2
Washington State University	4	15	Florida Institute of Technology	1	4
Claremont Graduate School/CA	4	16	Memphis, University of/TN	1	4
University of Chicago/IL	4	28	Mississippi State University	1	15
University of Toledo/OH	4	19	Northwestern State University/LA	1	8
University of Washington	4	32	Nova Southeastern University/FL	1	2
Wayne State University/MI	4	52	South Carolina State University	1	10
Florida Atlantic University	3	17	State U of New York-Binghamton	1	1
Harvard University/MA	3	26	U of Missouri-Kansas City	1	16
Kent State University/OH	3	22	University of Akron/OH	1	21
U of North Carolina-Greensboro	3	20	University of Arizona	1	30
University of Cincinnati/OH	3	23	University of Kentucky	1	21
University of Connecticut	3	43	University of Miami/FL	1	20
University of Mississippi	3	36	University of Montana	1	4
University of Rochester/NY	3	17	University of South Dakota	1	12
Western Michigan University	3	13	University of Wisconsin-Milwaukee	1	6
Arizona State University	2	31	University of Wyoming	1	12
Bowling Green State University/OH	2	3	Virginia Polytech Inst & State U	1	41
Clark University/MA	2	7	Ball State University	1	2
Emory University/GA	2	8	University of California-Davis	1	2
George Mason University/VA	2	7	University of Central Florida	0	10
Lehigh University/PA	2	11	TOTALS	1271	4157
Louisiana State U & A&M College	2	23			

Note: Data in the 2nd column are from NRC upon special request. Data in the last column are from *JRME* (July, 1980-1994) and DISS (1994-1997).

While size should not be the determining criterion in recognizing doctoral programs in mathematics education, it cannot be overlooked. The fact that over one-third of the institutions shown in table 3 produced a total of two graduates or fewer who declared mathematics education as their major field raises the question "What constitutes a program?" If only one or two graduates are produced over an 18-year period, can a program exist? No institution can afford to have regular graduate offerings without a cadre of students. In fact, most institutions of higher education require certain minimum course enrollments, and once the number of students becomes too low, regular course offerings cease to exist. When this happens, programs dissolve and are replaced by less well-defined efforts (often independent study) to provide the specialized preparation required for doctorates in specialty areas, such as mathematics education. This phenomenon is not limited to those institutions graduating one or two graduates but is experienced in different ways by nearly all institutions. Thus, in an effort to get a clearer picture of the scope of faculties, students, and doctoral programs in mathematics education, a survey was conducted.

CURRENT DOCTORAL PROGRAMS IN MATHEMATICS EDUCATION

A survey instrument was prepared to obtain information on mathematics education doctoral programs. Earlier survey instruments (McIntosh & Crosswhite, 1973; Batanero, Godino, Steiner, & Wenzelburger, 1992) were examined. Using this information, together with suggestions from the Organizing Committee for the National Conference on Doctoral Programs in Mathematics Education (see preface), a survey was designed to collect information providing different perspectives of mathematics education doctoral programs, including data about faculties, students, and programs. A paper copy of the survey was prepared along with an electronic version for the World Wide Web.

An announcement of the National Conference on Doctoral Programs in Mathematics Education and the availability of the survey was reported in several publications (*NCTM Newsletter,* 1999; *JRME,* 1999; *AMTE Newsletter,* 1999). An attempt was made to identify a mathematics education faculty member at each institution that awarded at least two doctorates in mathematics education from 1980 to 1996 (NRC, 1998). It should be noted that a limitation of this process is that it might have omitted some emerging programs in mathematics education. Locating a faculty member with expertise in mathematics education was not easy and often required personal contact (via e-mail or telephone) with department chairs or deans. Once an appropriate person was identified a letter and accompanying survey were sent to them.

Only one faculty member from each institution was asked to complete the survey (a time-consuming task). Since the survey was designed to provide institutional data, we felt it was better to request that data from one faculty member than to duplicate much of the same information by asking all faculty members to complete the form. Consequently, the survey data reflect a single individual's interpretation of his or her institution's doctoral program, and the extent to which this introduces bias is a limitation of this survey.

Surveys were mailed in March 1999. A survey form and cover letter describing the conference and the need for the survey data were mailed to a faculty member at each of the 81 institutions that had at least two graduates from the period of 1980 to 1996 according to the NRC's *Summary Report 1996.* In an effort to increase the rate of return, a follow-up letter was sent in April. E-mail and telephone contacts were made later to a faculty member at every institution with at least eight graduates from the period of 1980 to 1996 that did not return the survey from the first or second mailings. From the 81 institutions that were invited to respond, nine surveys were completed electronically and 39 were returned in paper format. In addition, four institutions (Northwestern University, University of Oregon, Harvard University, and Clark University) reported they currently had no doctoral programs in mathematics education.

Of the 48 returned surveys, 14 of the 15 programs identified as producing the most doctorates in mathematics education from 1980 to 1997, and 26 of the largest 30 programs responded. Thus, while the overall response rate was 64%, there was almost a 90% return from the largest 30 programs (i.e. Group 1 and 2 institutions as shown in table 5). These Group 1 and Group 2 institutions produced nearly 75% of the doctorates in mathematics education from 1980 to 1997 (NRC, 1999). Institutions that returned completed surveys are denoted by bold print in table 3.

In order to examine differences in programs according to the number of graduates produced, the 48 responding institutions were divided into three groups as identified in table 5. These groups of institutions will often be referred to throughout this paper.

It is important to note that some returned surveys had incomplete data (for example, missing information on number of males and females or ethnicity of students). The need to rely on incomplete data represents another limitation of this report. In such situations, we have typically cautioned the reader by indicating this as the 'percent of those reporting.' Selected information will be reported about faculties, students, and programs in the sections that follow.

Table 5. Institutions Completing Surveys by Number of Degrees Produced (NRC, 1999)

Group 1—Awarding at least 25 doctorates (1980–1997)		
Teachers College–Columbia University	New York University	University of Iowa
University of Georgia	Florida State University	Indiana University–Bloomington
University of Texas–Austin	University of Maryland	University of Wisconsin–Madison
Ohio State University	Temple University	Boston University
Georgia State University	Rutgers University	
Group 1 produced 662 of the 1271 or 52% of the doctorates in mathematics education from 1980 to 1997.		

Group 2—Awarding at least 8 but no more than 24 doctorates (1980–1997)		
University of Pittsburgh	Peabody College–Vanderbilt University	Illinois State University
American University	University of Tenessee–Knoxville	Stanford University
University of Oklahoma	Oregon State University	Oklahoma State University
North Carolina State University	University of South Florida	Southern Illinois University
University of Missouri–Columbia	Purdue University	Pennsylvania State University
University of Minnesota	Syracuse University	University of Michigan
University of Illinois	Auburn University	
Group 2 produced 275 of the 1271 or 22% of the doctorates in mathematics education from 1980 to 1997.		

Group 3—Awarding no more than 7 doctorates (1980–1997)		
Texas A&M University	University of Chicago	George Mason University
University of Alabama	University of Cincinnati	State University of New York–Albany
University of Delaware	University of Connecticut	University of North Dakota
Kansas State University	University of Mississippi	University of Central Florida
University of Colorado	University of Rochester	
Group 3 produced 48 of the 1271 or 4% of the doctorates in mathematics education from 1980 to 1997.		

RESULTS OF SURVEY

INFORMATION ABOUT FACULTY

Survey respondents were asked to list the name of each faculty member, areas of expertise, and institution and date when highest degree was earned. At the 48 institutions that returned surveys indicating they had a doctoral program in mathematics education, faculty size ranged from 1 to 16, with the mean number of faculty being five and the median being four. The distribution of faculty within Group 1, 2, and 3 institutions is provided in table 6. While some institutions reported only mathematics education faculty, others included faculty in areas such as psychology or mathematics. It should be noted that table 6 contains all the information reported by the person completing the survey and no screening of data occurred.

Table 6. Faculty Distribution of 48 Programs Responding to the Survey

	Total Number of Faculty	Mean Number of Faculty	Median Number of Faculty	Range of Faculty Size	% of Male Faculty	% of Female Faculty
Group 1	86	6	5	2–14	60%	40%
Group 2	93	5	4	1–16	53%	47%
Group 3	45	3	3	1–8	60%	40%
Total	224	5	4	1–16	57%	43%

Over one-third of the faculty listed in the responses graduated from the University of Wisconsin, the University of Georgia, Indiana University, Ohio State University, Teachers College-Columbia University, or the University of Maryland.

In addition, data on the year of highest degree earned yielded an interesting finding with regard to the age of current faculty. Of the 224 faculty listed, 211 (94%) also indicated the year when their highest degree was earned. From the information given in figure 1, it would appear that the average age of current faculty is fairly young since more than one-half graduated in the past twenty years. Examination of survey data on retirement however, revealed a different perspective.

Figure 1. Percent of Faculty Doctorates Earned by Decade

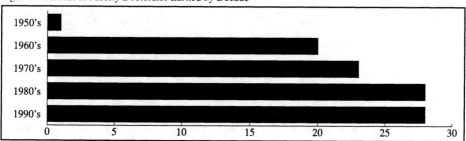

A follow-up survey question asked, "How many mathematics education faculty are *eligible* for retirement within 0-2 years? 3-5 years? 6-10 years?" Data indicated that 115 faculty of the 224 listed will be eligible for retirement within 0-2 years, 37 within 3-5 years, and 29 within 6-10 years (see figure 2). Thus, almost 80% of current faculty are eligible for retirement within the next 10 years. Consequently, although most doctorates were received in the last two decades, many of these were earned by people now nearing retirement. Other survey responses indicated that 38 of the 48 institutions anticipate hiring a collective total of approximately 75 faculty members in mathematics education within the next five years.

Figure 2. Percent of Current Faculty by Years Until Elligible for Retirement

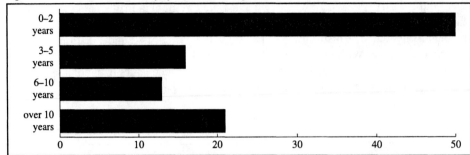

Other survey questions investigated respondents' perceptions of the current and future supply and demand of mathematics education faculty. As shown in figure 3, 78% of the respondents believe that currently there are more mathematics education jobs than qualified candidates, and 85% gave the same response when asked to forecast (looking ahead 5 to 10 years) future supply and demand for doctorates in mathematics education.

Figure 3. Rating of Supply and Demand for Doctorates in Mathematics Education

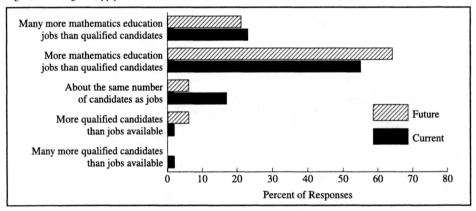

INFORMATION ABOUT DOCTORAL STUDENTS

The survey responses provided by the 48 programs in the survey provide a picture of the student population in mathematics education doctoral programs. The programs in the survey currently have a total of 734 students (332 full-time and 402 part-time). The highest number of students in any one program is 64 (30 full-time and 34 part-time), and the lowest is 0. Group 1 programs account for 367 (50%) of the 734 students; Group 2 programs, 256 (35%), and Group 3 programs, 111 (15%). Table 7 gives a breakdown of the enrollment status, gender, and ethnicity of the students for each group of programs. It is interesting to note that the larger programs tend to have a lower percentage of female students and a higher percentage of non-white students.

Table 7. Percent of Current Graduate Students in Mathematics Education by Demographic Category

Category	Group 1	Group 2	Group 3	Total
Full-Time	45	48	40	45
Part-Time	55	52	60	55
Male	41	40	23	38
Female	59	60	77	62
American Indian / Alaskan	0	0	6	1
African American	15	14	6	12
Asian American	4	1	3	3
Hispanic American	4	2	4	3
International	18	14	5	15
White (non-Hispanic)	59	68	76	66

The survey data show that financial support of full-time doctoral students comes primarily from institutional funds such as teaching assistantships. Approximately 53% of the doctoral students' support come from institutional funds, 26% from external grants, 16% from fellowships or scholarships, and 3% from other sources. The most significant difference in funding among the three groups of programs was that the larger programs reported a lower percentage of student support from institutional funds (Group 1, 42%; Group 2, 55%; Group 3, 62%), but a higher percentage of support from external grants, as well as fellowships, scholarships, or other sources (frequently the students' outside resources).

The median minimum financial support reported was $8,775 while the median maximum financial support reported was $13,600. The lowest level of support reported was $0 and the highest was $32,000. Examination of support by groups shows that the median minimum and maximum reported by Group 1 programs was $8,775 and $12,500, by Group 2 programs was $9,950 and $14,500, and by Group 3 programs was $6,500 and $11,000.

Apparently, survey respondents found the questions concerning recent graduates the most difficult to answer. The survey asked for the total number of graduates in 1998, in 1997, and from 1992 to 1996. The total number of graduates in 1998 was 97 as reported by institutions. In the *Summary Reports* (NRC, 1998; 1999), the 48 programs represented in the survey granted approximately 70% of all doctorates in mathematics education in 1996 and 1997. Using this percentage, we get an estimated total of 139 doctorates granted in 1998. That would represent a substantial increase in the yearly totals reported by the NRC (the highest number previously reported in table 3 is 115 in 1998). This increase could indicate either an underreporting by the NRC of the number of doctorates granted in mathematics education or an emerging upward trend. It could also mean that respondents to the survey had difficulty pinpointing the year of graduation for individual students and counted graduates who had actually received their doctorates in a different year. This interpretation might also explain why 34 of the respondents did not give a total number of graduates for 1997. For 1992 to 1996, most respondents (all but six) gave a total of 336 graduates, which was reasonably close to the *Summary Reports* total of 397 for the same period.

The survey also asked for the names and first jobs of students who graduated in 1997 and 1998. The 48 programs identified 203 graduates by name. It should be noted that some schools, in particular one of the largest, did not provide names for various reasons, including the students' right to privacy. Of the 203 graduates listed, 194 were identified with first jobs. We categorized these jobs into the following eight groups:

U.S. universities with doctoral programs in mathematics education (represented in table 3), U.S. universities without doctoral programs in mathematics education, U.S. community or junior colleges, U.S. K–12 schools, U.S. commercial or governmental agencies, international universities, international governmental agencies, and unknown. The percentages for each of these categories are found in figure 4.

Figure 4. Percent of 203 Mathematics Education Doctorates in 1997–1998 by First Job

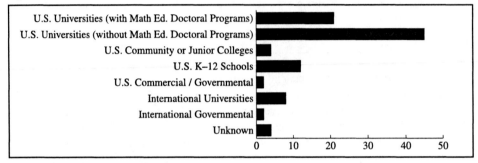

These totals are consistent with the way programs ranked those jobs taken most frequently by their graduates. Table 8 shows the possible jobs listed on the survey and the number of first or second rankings received for each possibility. Both figure 4 and table 8 show the predominance of higher education jobs taken by graduates, but figure 4 shows more clearly that the largest percentage of graduates (45%) are first employed at U.S. universities without mathematics education doctoral programs.

Table 8. Frequency of First and Second Rankings of Positions Taken by Graduates in Mathematics Education

Position	Ranked First	Ranked Second
Classroom Teacher in K–12 Schools	3	2
Mathematics Coordinator / Supervisor	2	2
Higher Education–Mathematics Dept.	19	10
Higher Education–Education School	19	14
Higher Education–Joint Appointment	3	12
Commercial Companies	0	0
Researcher at Non–University Institute	0	0
Other	2	0

Figure 5 shows the number of reported graduates within each institutional group who took first jobs in each category. As already noted, the most frequent first job of graduates of doctoral programs in mathematics education is at a U.S. university without a doctoral program in mathematics education, and it is interesting that this is true even for graduates from Group 1 programs. It should also be noted that Groups 1 and 2 produce nearly all of the graduates who have first jobs in the international community, while Group 3 programs are more likely to produce graduates that take first jobs at U.S. K–12 schools. (13% of Group 1 graduates, 7% of Group 2 graduates, and 15% of Group 3 graduates take first jobs at U.S. K–12 schools).

Figure 5. Number of Reported Mathematics Education Doctorates Completed in 1992–1998 in Each Group by First Job

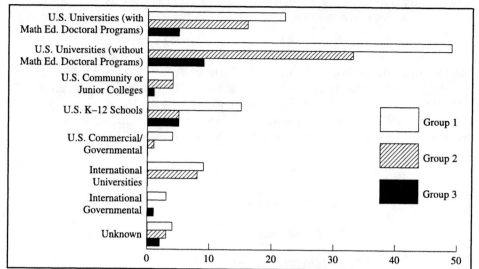

INFORMATION ABOUT PROGRAMS

Thirteen of the fourteen responding doctoral programs in Group 1 have been in existence for more than 30 years. The exception, Georgia State University, is in the 10 to 30 year category. Approximately 55% of the programs in Groups 2 and 3 have also been in existence for over 30 years, but their production of doctoral candidates is considerably less than that of Group 1.

The degree offered at most of the institutions that responded is the doctor of philosophy (Ph. D.) with 92% of the respondents granting this degree. The doctor of education degree (Ed. D.) is offered at 44% of the responding institutions, with Group 1 having the highest percentage (50%). The doctor of arts degree (D. A.) is not offered at any of the institutions. These doctoral programs in mathematics education are housed in numerous departments and colleges. Over two-thirds of the programs are housed in their institutions' school or college of education. Most of the remaining programs were housed in a school or college that combined education and either psychology, human development, or arts and science.

The doctoral programs are also housed in various departments within their respective schools or colleges, with Curriculum and Instruction being the most prevalent. There are only two departments of mathematics education, one at the University of Georgia and the other at University of Texas at Austin. Two universities (Illinois State University and American University) reported their doctoral programs were housed in the College of Arts and Science and situated in the mathematics department.

For students who wish to enter a doctoral degree program, there are minimum requirements at all reporting universities, although an institution's requirements differ from a department's requirements. The requirements reported most often at the department level are minimum GRE scores and minimum GPA, required in 75% and 71% of the responding universities, respectively. These are followed by three requirements: K–12 teaching experience (56% of responses), qualifying exams (54% of responses), and a prior BS/BA degree in mathematics (48% of responses). The largest program, Teachers College, cited only two departmental requirements: K–12 teaching experience and a qualifying exam.

There were differences in entrance requirements among the programs in the three groups (see figure 6). At the department level, the most common requirements for entering programs in Group 1 are minimum GRE (79%), minimum GPA (71%), and qualifying exams (71%). In Group 2 programs, prevalent requirements are minimum GRE (85%), minimum GPA (75%), English proficiency (60%), and bachelors in mathematics (60%). In Group 3 programs, the most common requirements are K–12 teaching experience (71%), minimum GPA (64%) and minimum GRE (57%).

Figure 6. Percent of Departments in Each Group Having Each Type of Entrance Requeirment

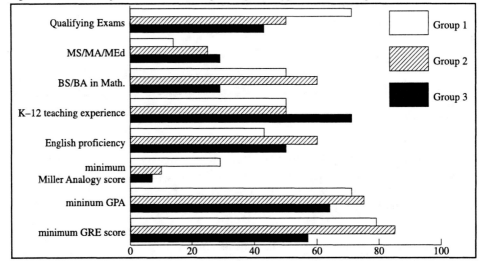

At the institutional level, there were no major differences among programs of the three groups, with the two most prevalent requirements being minimum GPA and English proficiency (64% and 62% of responses, respectively). The Other category in both the departmental and institutional requirements included a mixture of individualized requirements, such as writing samples from the applicants or portfolios. It is worth noting that departmental requirements tend to be greater than those at the institutional level.

Most schools require residency (85%), comprehensive exams (96%), a dissertation (100%) and a defense of the dissertation (98%) from their doctoral candidates. Among the programs, one from Group 1, three from Group 2, and two from Group 3 also require a presentation of dissertation findings at a conference (see table 9). The number of graduate semester hours needed to complete a doctoral program ranges from 45 and 125, with wide disparity. Five programs listed 50 semester hours or fewer as the typical number needed to complete a doctorate. Of these five, the University of Illinois has a master's degree entrance requirement, and the University of Connecticut stated that the 45 semester hours needed to complete the doctorate is in addition to the hours required for the master's degree. The other three programs did not explain their reported number of semester hours.

Table 9. Percent of Programs by Group that have Each Graduation Requirement

Group	Residence	Exam	Dissertation	Oral Defense	Published Article	Presentation at Conference	Grant Proposal
1	79	100	100	100	0	7	0
2	80	95	100	100	0	15	0
3	100	93	100	93	7	14	0

Even though most doctoral programs are individually tailored to reflect each candidate's background and interest, there are some mandatory areas in any program of study where candidates are expected to obtain a depth of knowledge. Respondents were asked to rate the emphasis on these knowledge areas using a scale of 0 to 3, with 0 reflecting no emphasis and 3 indicating major emphasis (see figure 7). What may be surprising about these responses is the relatively low emphasis placed on technology (average of 1.8), especially in Group 3, and general foundations (average of 1.9). Across the three groups, the area with the greatest difference in emphasis is research methods. It is particularly interesting that Group 3 programs report a major emphasis on research in mathematics education but only a moderate emphasis on research methods. As for mathematics content knowledge, this area receives, on average, a moderate emphasis, yet most programs assume a strong mathematics requirement for acceptance into the doctoral program.

Figure 7. Average Emphasis Given Within Program by Group

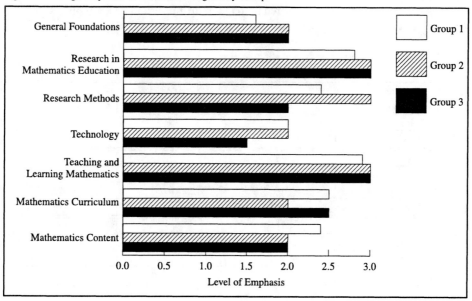

Figure 8. Percent of Institutions in Each Group Requriring Candidates to Attain Each Level
in Mathematics Coursework

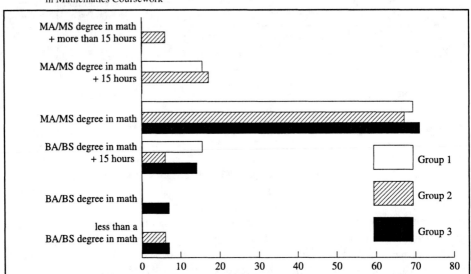

As figure 8 shows, over 60% of the respondents stated that their successful doctoral
candidates will have attained a master's level in mathematics. A wider range of
competence in mathematics is required among Group 2 programs than in programs in
Groups 1 or Group 3.

Table 10. Types of Dissertations, in Percent of Responses

Group		Quantitative	Qualitative	Pure Mathematics
1	Preferred	29	43	0
	Acceptable	71	57	7
	Discouraged	0	0	71
	Unacceptable	0	0	21
2	Preferred	35	25	0
	Acceptable	65	75	5
	Discouraged	0	0	79
	Unacceptable	0	0	16
3	Preferred	31	15	0
	Acceptable	69	85	9
	Discouraged	0	0	64
	Unacceptable	0	0	27

With dissertations being required in all 48 programs studied, the survey also asked
which types of dissertations are acceptable (see table 10), as well as which types are most
frequently submitted by doctoral candidates. Qualitative and quantitative dissertations
were considered acceptable by all those who responded to this question. Qualitative
dissertations, however, were preferred by a greater percentage of programs in Group
1, while a greater percentage of programs in Groups 2 and 3 preferred quantitative
dissertations. The dissertation research design most frequently submitted by doctoral

candidates is fairly evenly split between qualitative, quantitative, and a mixture of these two types.

The remainder of the survey invited respondents to give more insight into individual characteristics of their program. First, respondents were asked to describe any unique features of their programs. While the responses varied, the main distinguishing features included: (1) programs designed around varied career goals and needs of individual students, (2) apprenticeship or active participation in research and other projects, (3) opportunities to work closely and collaboratively with faculty, (4) depth of study of mathematics, and (5) opportunities to teach (co-teach) methods courses or work with teacher interns.

Another question asked for any special areas of emphasis in the doctoral programs. Special areas of emphasis included: (1) content preparation, (2) elementary teacher education and pre-service and novice teachers' beliefs, (3) middle school teachers' professional development, (4) equity and diversity issues, (5) urban education, and (6) preparation of doctoral students to teach in college mathematics departments or in higher education and research.

Respondents were also asked to address recent changes (those in the last five years) and anticipated changes (projected for the next five years) within their programs. Twenty-seven of 47 indicated that their programs have changed in the last five years. Recent changes included more emphasis on use of appropriate technology, addition of option for concentration in middle school mathematics education, and more emphasis in graduate research and qualitative methodology. More changes are anticipated by 36 of the 46 respondents. The majority of anticipated changes pertained to an increased emphasis in technology and the use of distance learning in their programs. Other predicted changes included the recruiting of more minority students and international students, new or increased focus on middle school education, and preparation for college teaching. One respondent summarized by stating, "It is hard to imagine any program remaining unchanged. As faculty interests evolve and students change, new program emphasis is certain to evolve. The nature of these changes is not certain, which is what makes the future so exciting."

SUMMARY

The survey information regarding the 48 institutions provides a current picture of doctoral programs in mathematics education in the United States. The picture is composed of information on faculty and students, as well as particular characteristics of individual doctoral programs in mathematics education. This study led to a related investigation of recent graduates of doctoral programs in mathematics education (Glasgow, 2000).

Examining the information about mathematics education faculty of these doctoral programs revealed several interesting findings. For example, the number of mathematics education faculty in a department ranged greatly (from 1 to 16) and the majority of these faculty members had earned their degrees since 1980. The most surprising finding was that one-half of the current mathematics education faculty members are eligible for retirement within the next two years. Furthermore, over three-fourths of current faculty are eligible to retire within the next ten years. Since there is already a shortage of doctorates in mathematics education in the United States, the significant increase in projected retirements promises to enlarge the current problem. Furthermore, the rapid "changing of the guard" among mathematics education faculty in doctoral-granting institutions is likely to change significantly the nature and structure of many of these doctoral programs.

The number of doctoral students in individual programs varied from 0 to 64. A majority of these students are white, as well as female. The full-time doctoral students receive, on average, $10,000 in annual support. Most of the graduates from these programs take their first jobs in higher education, with the majority of these higher education positions being at institutions without a doctoral program in mathematics education. Graduates from larger institutions tend to take positions outside of the United States more than do those from smaller programs; graduates of smaller programs are more likely to take positions at K–12 schools and community colleges.

Although doctoral programs offering a degree emphasizing mathematics education have unique characteristics, they also share many features. The majority of universities reported minimum requirements in GRE scores and grade point averages for entering doctoral students. Upon completion of a program, the majority of universities also expect graduating students to have a strong background in mathematics content, mathematics education research, and in pedagogy and learning theory. While some universities do not have an examination requirement for graduating students, all have a dissertation requirement, and most require an oral defense of the dissertation.

This survey, together with information reported by the NRC, *JRME*, and DISS, provide much food for thought and reflection regarding the production of doctorates in mathematics education and the nature of their doctoral programs. The survey highlights the difficulties of obtaining timely, accurate data about doctoral programs while underscoring the regular need for such data. Of course, our data provide only a snapshot at a point in time, and we realize they are incomplete as well as dated, given how quickly programs change. The survey, however, does provide benchmark data that suggest critical issues in need of careful and thoughtful attention. Any significant legacy of this effort will be measured by the ways in which these survey data are used for discussion and, ultimately, action directed toward better defining and improving doctoral programs in mathematics education.

Robert E. Reys
121 Townsend Hall
University of Missouri
Columbia MO 65211
reysr@missouri.edu

Bob Glasgow
Department of Mathematics
Southwest Baptist University
Bolivar MO 65613
bglasgow@sbuniv.edu

Gay A. Ragan
School of Teacher Education
Southwest Missouri State University
Springfield MO 65804
gar098f@smsu.edu

Kenneth W. Simms
Hickman High School
1104 N. Providence
Columbia MO 65201
ksimms@hhs.columbia.k12.mo.us

CBMS Issues in Mathematics Education
Volume 9, 2001

REFLECTIONS ON THE MATCH BETWEEN JOBS AND DOCTORAL PROGRAMS IN MATHEMATICS EDUCATION

Francis (Skip) Fennell, Western Maryland College
Diane Briars, Pittsburgh Public Schools
Terry Crites, Northern Arizona University
Susan Gay, University of Kansas
Harry Tunis, National Council of Teachers of Mathematics

This presentation and the discussion that followed dealt with reflections of the participant panel on their doctoral programs and the extent to which they prepared them for their roles in mathematics education. Panelists discussed their varied experiences, their current positions, and how their backgrounds have been shaped by their doctoral studies. Participants offered suggestions related to perceived needs in doctoral programs. This paper is organized according to three major elements, namely scholarship, teaching and service. Interestingly, these elements are also critical components in the retention and promotion of college faculty nationwide.

SCHOLARSHIP

All of the panelists recognized the need for doctoral programs to advocate scholarship; however, the use of the term scholarship differed. Most of the panelists discussed the importance of mathematics content in the background of prospective mathematics educators. Some felt that more mathematics was needed. Most programs seem to require the equivalent of an M.A. in mathematics. This requirement seems most appropriate, or perhaps more realistic, for candidates who had an undergraduate major in mathematics, but it may eliminate elementary and middle-school teachers from candidacy.

Panelists also noted that in some ways the mathematics learning component of their programs was torturous, a comment which engendered more discussion than any other aspect of the panel presentation. Completing rigorous mathematics course work while fulfilling the other duties of a graduate student was seen as a challenge, but panelists and members of the audience considered mathematics course work an important component of any doctoral program in mathematics education. Ideally, mathematics experiences should reflect the most appropriate pedagogy, including the use of technology.

Perhaps there should be some forum in which graduate students could discuss with their mathematics professors the thinking that goes into the development and teaching of mathematics courses. Susan Gay's program included course work in four areas of mathematics: analysis, algebra, geometry, and probability and statistics. Although Terry Crites had a master's degree with emphasis in mathematics, he felt that he would have benefited from more graduate-level course work in mathematics. However, panelists generally felt their programs had been more concerned with the mathematics background of their candidates than their knowledge of and experience within the Pre-K–12 classroom. That is, most programs did not require teaching experience, but all required mathematics. The existence of such a difference relative to mathematical content knowledge and classroom experiences within mathematics education programs may be an issue for further discussion.

While panelists discussed the role of mathematics in their doctoral programs, there was little discussion of specific course work in mathematics education, instruction, curriculum, and learning or of related course work and experience in research. Harry Tunis's reference to scholarship focused on the need for doctoral candidates to read widely; such reading should include editorials, articles on mathematics and mathematics education, and research. He felt the opportunity to critique and discuss such work should be an integral component of any doctoral program. Terry Crites indicated that he wished he had more opportunity to listen to his professors "think aloud" about how to organize, conduct, and analyze research. Skip Fennell noted that a critical component of any graduate program is active involvement in a research agenda. Susan Gay noted that doctoral students need to understand a wide range of research paradigms, from classic experimental research design to qualitative models. Students should read and be able to synthesize such work. This issue led to some discussion of the role of the dissertation within the doctoral program. Over the years, this program capstone has received its share of emphasis and criticism. Regardless of the future role of the dissertation, conference participants acknowledged the value of research and writing. The panel agreed that doctoral programs should provide opportunities for students to write. It was noted that the significant professional contributions of many mathematics educators are related to their ability to write. Doctoral students must develop as writers within the profession and be aware of the submission guidelines for professional journals that shape our field—be they journals for practitioners, mathematicians, mathematics educators, or researchers.

TEACHING

Skip Fennell mentioned the need for doctoral programs to provide opportunities for teaching. Doctoral students should experience teaching at the Pre-K–12 levels. Programs that do not offer such opportunities should list them as prerequisites. Programs may emphasize certain levels of mathematics education (i.e., early childhood, elementary, middle, secondary), and if they do so, teaching and/or practicum experiences should be required. Additionally, doctoral students should have guided experience in teaching college-level mathematics and/or mathematics education classes. All too often the ineffective model of "throwing such neophytes to the wolves" has been the norm. Harry Tunis recalled a team-teaching experience with fellow professors (after his doctoral program), in which he learned a great deal from the joint planning and delivery of instruction. Terry Crites indicated that his doctoral program experience with various mathematics education courses and the supervision of student teachers was an appropriate way for him to gain pedagogical techniques. However, he also noted that doctoral students who do not have experience teaching at appropriate grade levels will have little credibility teaching (pre-service) mathematics teachers or leading staff development. Such students should be advised to acquire teaching experience. Susan Gay has her doctoral students supervise student teachers and, when possible, teach mathematics courses for elementary teachers. She guides individual graduate students by instructing one section of a course and letting the student teach a second section. She notes that this modeling approach is particularly helpful for doctoral students who have never taught "real" elementary students. On the other hand, Diane Briars has been a very successful coordinator/supervisor of mathematics in Pittsburgh, Pennsylvania, for many years but has only limited classroom teaching experience.

The panel opinion was that teaching is an important component of a doctoral program in mathematics education. Most doctoral programs involve graduate students in teaching mathematics or mathematics education courses. Such "teachers," like novices in any field, need assistance in developing course syllabi, expectations and goals, and strategies for instruction within the courses. Nurturing is needed here. If the profession values teaching, it must provide for an apprenticeship in this area that will vary with the success of the candidate. Likewise, for those students contemplating careers in teacher education, whether it be preparing teachers or providing staff development, knowledge of and experience with students at the Pre-K–12 levels are essential. Thus, any doctoral program designed to prepare those students must include a teaching component that reflects a mentoring model.

SERVICE

Doctoral students should become involved in their profession. They should be encouraged to take active roles in organizations affiliated with mathematics and mathematics education. Such opportunities could include speaking at state mathematics conferences, reviewing and writing for state and regionally published mathematics education journals, and/or serving as referees for National Council of Teachers of Mathematics (NCTM) journals. Skip Fennell noted that doctoral students may begin to serve as consultants during their graduate student days. Such opportunities must be carefully developed, and students need mentoring and nurturing as they take on such activities, which may involve, among other things, writing or reviewing curricula, providing professional development, or serving as project evaluators.

Educators who design doctoral programs should seriously consider developing courses and/or related program experiences to assist doctoral candidates in the area of leadership. Topics could include: examining how state and local education decisions are made, particularly as they affect mathematics teaching and learning; exploring consultant roles open to mathematics educators; and exploring the availability of external sources of funding, particularly those presented by the National Science Foundation, the United States Department of Education, and private foundations. Discussion should underscore the critical role technical writing plays in all these areas, especially in the funding process. An important added component would be to have students translate their experiences into articles for scholarly journals.

NEXT STEPS

This conference was very timely. With student numbers dropping, there is a tremendous need for the next "generation" of mathematics educators. Interestingly, much of the discussion at this session and throughout the conference seemed to center on doctoral programs that remain remarkably similar. Parallel programming during a critical shortage does not meet the needs of the profession. In fact, greater diversity among programs is needed. For example, Diane Briars questioned many programs' maintaining the residency requirement. She and others discussed the financial and personal burdens that such a requirement places on teachers. Some conference participants discussed possible alternatives to the dissertation. Others suggested doctoral practicums within state departments of education, the United States Department of Education, and the National Science Foundation. Several of the panelists suggested that doctoral programs provide some career guidance for their graduates; paths may include appointments at institutions of higher education, ranging from small, private liberal arts institutions to large, state-supported universities. Other graduates may work in administrative or supervisory positions in school districts. Still others may seek positions in foundations, governmental agencies, publishing, and so on. To what extent should doctoral programs make their graduates aware of such opportunities? This question and the others raised in this paper point to a tremendous need to examine and reevaluate the elements of doctoral programs in mathematics education, taking into consideration the diminishing pool of graduate students as well as the unique needs within our profession.

This panel session provided a glimpse of lessons that participants had learned along the way—from doctoral program involvement (as students), through current careers in the field. The panelists are all deeply invested in the future of mathematics education. Doctoral program graduates represent the future; they must be among the leaders in guiding efforts to improve the quality of mathematics education at all levels.

Francis (Skip) Fennell
Western Maryland College
Westminster, MD 21157
ffennell@wmdc.edu

Diane Briars
Coordinator of Mathematics
Pittsburgh Public Schools
150 Maple Hts Rd
Pittsburgh, PA 15232
briars@pps.pgh.pa.us

Terry Crites
Department of Mathematics and Statistics
Northern Arizona University
Box 5717
Flagstaff, AZ 86011
terry.crites@nau.edu

Susan Gay
Department of Learning/Teaching
and Mathematics
University of Kansas
202 Bailey Hall
Lawrence, KS 66045
sgay@ukans.edu

Harry Tunis
Director of Publications
NCTM
1906 Association Drive
Reston, VA 2091
htunis@nctm.org

CBMS Issues in Mathematics Education
Volume 9, 2001

International Perspectives on Doctoral Studies in Mathematics Education

Alan J. Bishop, Monash University

Introduction

This is an important conference on doctoral programs in mathematics education and it is being held at this most opportune time. With global communications increasing at a rapid pace, with international travel and conferences now being commonplace for every academic, and with the new millennium upon us, I believe this is a good time to consider the international dimension of doctoral studies in our field.

However, preparing this talk has been a challenging, even daunting task, since I cannot be expected to know what is going on everywhere in the world. From my inquiries it seems that the topic could easily be worked into a good Ph.D. dissertation if anyone is still looking for such a topic! With current interests in social, cultural, and political aspects of our field becoming increasingly prominent, there is no shortage of worthwhile international issues and problems to address.

All I can do in this short time and space is raise what I think are some interesting questions and provide partial answers from my own experience and that of a few colleagues whom I have contacted. For myself, when I finished my undergraduate degree and preservice teacher training in the U.K., I was advised by my tutor at Southampton University, Bill Cockcroft, to try for a scholarship to do graduate study at Harvard graduate school of education. This I did, was successful, and spent three fascinating years getting my MAT and doing most of the coursework for an Ed.D. Sadly, the money ran out, and family commitments grew, so I returned to the U.K., and there I duly completed my Ph.D. at Hull University. Since then I have supervised doctoral students at Cambridge University and at Monash University in Australia, as well as acted as external examiner for many universities in other parts of the world.

I do, therefore, have a variety of experiences to draw on for this talk, but I cannot claim to know precisely what is going on in every country. I claim no sweeping generalisations from my "data"; indeed, I cannot even guarantee that the data are 100 percent accurate! However, if the paper provides some challenges to your thinking and also lays down some interesting avenues that could be explored in later research, then I will be satisfied. I just hope that my international colleagues will forgive my possible blunders about their countries' procedures.

TO WHAT EXTENT IS DOCTORAL STUDY IN MATHEMATICS EDUCATION AN INTERNATIONAL PHENOMENON?

There is no doubt in my mind that it is an international phenomenon. Doctoral studies in mathematics education take place in several countries, though by no means all. These studies exist at least in the U.K., France, Canada, Germany, Australia, Denmark, Ireland, Holland, Belgium, Israel, Singapore, New Zealand, Taiwan, South Africa, Spain, and Malaysia, to name the ones with which I am familiar. However, research students live in a far greater range of countries, as many students study overseas, either by preference or because they have no opportunity to study for the degree in their own countries.

The studies take different forms, although similarities can be found, and more will be said about them below. For example, the research topics are meaningful internationally, as we can see at international and regional conferences, and are not restricted in meaning to one country. The research methods used are also meaningful internationally, although what those methods are is now an area of strong concern and interest, as we shall see below.

There is a great deal of international co-operation in doctoral work through the use of references, international conference papers, the selection of external examiners, and the giving and receiving of advice from many quarters. The international publication of doctoral research is increasing also, as well as distance education (see various chapters of the *International Handbook*, Bishop et al., 1996). These developments are increasingly focusing attention on the international dimension of doctoral research.

However, doctoral study around the world may be characterised as a neo-colonial phenomenon (and I include the United States as a neo-colonial power for the purposes of this paper). For example, most doctoral studies are done in an "international" language, the majority being in English, German, or French. Many international students do their studies in their former colonial rulers' countries, where they can get access to important books, facilities, and so on. Indeed, it is often the case that appropriate materials about the student's home country may only be found in the former colonial ruler's libraries. Financial support in the form of scholarships and grants is often given by the wealthy colonial countries from their overseas aid budgets, and it is often targeted on their former colonies.

Thus, doctoral study in mathematics education is not a non-political, value-neutral activity; it is wide open to diplomatic, economic, and political pressures. Its development is subject to economic and market forces both internally, within individual countries, and externally, within the international community. For example, international students are seen as valuable sources of income for universities struggling to maintain student numbers in the face of declining local interest. Much money and time is invested in recruiting and nurturing them. Indeed there are parallels here with another development in the United States, where mathematics departments are increasingly offering their own Ph.D.'s in mathematics education. In part this is a response to their falling enrollments as students, and often the best ones, choose to go into technology or business faculties.

WHAT DOES "TO TRAIN TO DO RESEARCH IN MATHEMATICS EDUCATION" MEAN IN OTHER COUNTRIES?

Unlike in the United States, where doctoral studies are seen as a way of training students for a wide variety of positions in the educational system, internationally the emphasis of doctoral study is rather on developing staff for university departments. Thus the word, and the idea, of research "training" is an important one to consider.

There are at least two ways to think about answers to this question:

- What does the word *train* mean internationally? Is it a viable concept?
- What does the word *research* mean internationally?

Regarding the first, although there is an element of research training in all doctoral study, there is wide variation in the ways this is interpreted internationally. Are there specific research-training sessions in mathematics education? In fact, in most countries the number of doctoral students in mathematics education at any one institution is generally quite small, and so it is neither economical nor productive to run specific research-training sessions in mathematics education. One takes training by sitting next to the expert, in the words of the old image! In this case the expert would be the research supervisor, and training would be defined almost as following in his or her footsteps.

I take training to do research into mathematics education to have a different goal from that of helping research students with their individual research projects. It seems also to assume that each student will become either a full-time researcher, expected to know and be familiar with a range of methods, or a full-time academic with supervisory responsibilities for a wide range of research students. It, therefore, also makes an assumption about the career aspirations of the average research student.

In most cases it certainly seems as if the main motive for students in other countries to do doctoral studies is to develop a career path toward a university career. However, this is not now always the case, as we can see with the Ed.D. degree, which is becoming more popular in the U.K. and Australia now. At Monash University, for example, the Ed.D. is for practising professionals in various aspects of the education service who do not always want to move into universities. I will say more about this later.

Regarding the second question, in my chapter in the *Handbook of Research on Mathematics Teaching and Learning* (Bishop, 1992), I outlined the ways in which research in our field is conceived differently around the world, basing my discussion on three fundamentally different educational traditions:

One can see differences, for example, between the Anglo-Saxon tradition (a combination of 1 and 2) and that on the continent of Europe (a combination of 1 and 3). This often affects the kind of research study that students in the different former colonies undertake.

It is the kind of difference that can also be seen between the research discussions at PME—International Group for the Psychology of Mathematics Education (traditions 1 and 2)—and at CIEAEM—Commission internationale pour l'étude et l'amélioration de l'enseignement des Mathématiques or the International Commission for the Study and Improvement of Mathematics Education—(traditions 1 and 3).

A particular corollary to this question is whether one can do a research study on mathematics education in an education department/faculty, or whether one must do it in a mathematics department/faculty or a psychology department/faculty. I say it this way, which I know sounds prejudiced, because the problems of territory are all about whether education exists as a discipline at university (as mathematics invariably does and

Tradition	Goal Of Inquiry	Role Of Evidence	Role Of Theory
1. Pedagogue tradition	Direct improvement of teaching	Providing selective and exemplary children's behaviour	Accumulated and shareable wisdom of expert teachers
2 Empirical-scientist tradition	Explanation of educational reality	Objective data, offering facts to be explained	Explanatory, tested against the data
3 Scholastic philosopher tradition	Establishment of rigorously argued theoretical position	Assumed to be known, otherwise remains to be developed	Idealized situation to which educational reality should aim

psychology often does), the answer to which determines whether it can house a Ph.D. program. This debate is still going on I believe in, for example, France, Italy, Greece, Spain, Portugal, and several other countries. Of course a student could, if he or she wished, pursue a doctoral degree in psychology in most of those countries with a leaning toward learning mathematics but not with an educational emphasis. Nor is it always the case that mathematics departments accept research studies in mathematics education. Even if they do accept the possibility, the educational scope of the study can be very limited (e.g., only to studies of mathematics curricula).

To What Extent Is Doctoral Study In Mathematics Education A "Taught" Study?

This question follows from the previous one. Is course work a component of all international doctoral study in mathematics education? No, it most certainly is not. In Australia, Germany, Denmark, the U.K., Ireland, and New Zealand, for example, it is not required. However the master's degree is generally a prerequisite for entry to doctoral study, and that degree will usually be a course work degree with courses on research methodology.

Sometimes the master's courses can be embedded in the doctoral study; for example, when an applicant has not reached the necessary standard there are often taught courses that can be taken for credit, or audited, before the student starts on his or her thesis work. Also there is rarely any required course work in mathematics. Indeed, where mathematical study is encouraged among the doctoral students, it is now being recommended that study of the history or the philosophy of mathematics would be preferable.

Usually, however, even if there are no formal course work requirements, there will be a research seminar group or two that the student can join. These often serve as social rather than purely academic supports; they might also contain students from other disciplines if the mathematics education numbers are not large. Doctoral students everywhere would have supervisors, or, as in the United States, even committees.

In some systems, the doctorate is usually looked upon as being the opportunity for some creative research work. I would offer a hypothesis here: the more the system employs detailed taught courses at this level, the less likely it is to accept, or even generate, creative Ph.D. dissertations! When I think of some of the brave, creative Ph.D. dissertations that I have been lucky to either supervise or examine, there has generally been little course work support, but the creative aspect has been fundamental, and that is not something that, to my mind, can be attained by training.

In Germany, and other German-speaking lands, they have the concept of a second doctorate, a higher doctorate, which is needed to teach at a university. So firstly, students do their regular doctorates as postgraduate work, which can take from 2 to 4 years. After that they start on what a colleague calls "much more voluminous and pretentious investigations and theoretical considerations that can easily take another 4-6 years. And a whole faculty will have to accept the product, a thick book usually." In Denmark they also have the concept of a higher doctorate, but there the system allows one thesis to be entered for either the higher doctorate or the ordinary Ph.D. This seems to encourage the production of some large as well as some brave, creative dissertations.

But there is also another way to consider this second question, and that is by focusing on the idea of a professional researcher. In many countries this is not a great career path, but where full-time research positions exist, they involve working in a ministry department, where research is seen as a data-gathering, or feasibility, exercise with limited theoretical aspects to it. In another case, I think only a few people are willing to tolerate the insecure occupation of contract researcher for any length of time. But, in some cases, the opportunity to do a postdoctoral fellowship for a few years enables a student to get some "runs on the board" in the university position stakes.

Another aspect to this question concerns the distinction between a Ph.D. and an Ed.D. The latter degree is very new in most countries and is only found in English- speaking countries or their colonial connections: for example, the United States, Canada, the U.K., South Africa, Australia, New Zealand, Singapore, Hong Kong, etc. One reason the education doctorate does not exist in other countries is because, as was mentioned earlier, education is not considered an academic discipline. Ph.D.'s will then, as in France, be done in mathematics or psychology departments/faculties.

In all cases where it does exist, the distinction is drawn between the specialist notion of the Ph.D. and the breadth of the Ed.D. "for practising educational professionals." The assumption made is that the latter are not seeking university employment, although all universities are eager to stress that the two doctorates are equivalent in stature, entrance requirements, and so on.

Is the distinction between the two meaningful? In the statements put out in the glossy brochures from the various departments, I have seen that there is certainly a big difference, for example, in the length of the dissertation and in the examination requirements. At Leeds University in the U.K., the Ph.D. requires a dissertation of 100,000 words, whereas for the Ed.D. there are 8 required courses, each with its own examination requirement, while the dissertation need only be 40,000 words. There is also, in many people's eyes, a status issue: the Ph.D. is seen as a higher-level degree. Thus, while the Ed.D. is in its relative infancy internationally, anyone thinking about pursuing a doctorate should consider carefully the implications of doing the Ed.D. It needs to gain credibility; at this point, it might be a handicap in applying for some university positions.

WHAT ARE THE MAIN CULTURAL ISSUES CONCERNING DOCTORAL STUDIES WITH INTERNATIONAL STUDENTS?

This question could certainly be the basis of a substantial study. Let us first take the issue of language. In what languages are their dissertations written? The vast majority are written in English, German, or French; also used frequently are Dutch, Spanish, or Japanese. For international students generally their dissertations would be in the language of the host country, which would commonly be their (neo-) colonial country.

In some cases, therefore, students can be writing in their third or even fourth language. At Monash I had two students from Papua New Guinea who each had a village language, a regional language, Pidgin, and English. Two other students from Mozambique were fluent in their village languages (and knew other Mozambican languages), Portuguese (the colonial language), German (the language in which they did their master's degrees), and English for their doctorates.

But this situation is not due just to the languages themselves, but to their position of power in mathematics education. The majority languages of mathematics education are English, French, and German. The references are in those languages, as are most of the books. International conferences are rarely linguistically international, and research students seeking to make their names on the world stage would get nowhere with dissertations in Swahili or Tamil. Their future job opportunities increase many times if they have done their doctorates in English and in English-speaking countries.

Another issue is whether the research done by a student is relevant to the home country's situation. Whenever I have an international research student I insist not only that the research issue be located in the student's home-country situation, but also that the main data gathering be carried out in the home country. I know that visa problems can sometimes make this difficult, even impossible, but the danger with not doing this is that the student educated completely away from his or her home culture may never return home.

Important issues are now being raised about the general applicability of research methods from one culture to another. The provocative paper by Valero and Vithal (1998) addresses this issue in a way that can inspire powerful debates that all of us involved in international study of mathematics education cannot ignore. It is yet another example of the neocolonialism of the mathematics education field. The research on ethnomathematics has sensitized us to the effects of neocolonialism on mathematics curricula in many countries, but so far few of us have begun to consider the cultural "loading" of research methods and approaches.

It may also be the case that a research approach or style is unacceptable "back home." I have heard stories from students who insist that they must carry out a certain kind of study, because that is the only kind their government and ministry of education will recognise. Doctoral study is hard enough at the best of times, without the additional complications of trying to reconcile the conflicting demands of supervisors, institution, and one's home government.

Finally, an appropriate research study for a local student may be totally inappropriate for an international student. He or she will be in a difficult situation not knowing the history or the culture of the situation. For example, one popular style of research for international students is the comparative kind, but the value of such a study is often undermined because the study will be done from the perspective of the host, rather than the home, country or situated in a totally different technological environment. It is better to use the immigrant or language experience of the student and to build positively on that in the Ph.D.

WILL DOCTORAL STUDY IN MATHEMATICS EDUCATION BECOME A MORE INTERNATIONAL ACTIVITY?

In my view, certainly it will. Doctoral research degrees are a Western development, and as the Western culture increases its dominance around the world, so more universities in other countries will want to have their own doctoral programs. Moreover, as the "Anglo" Western world continues to exert its dominance through language and media, so more and more future academics in other countries will want to be credentialed in English. ICMEs are now almost entirely in English as are several regional meetings (with perhaps the local language accepted also), and of course the major research journals and books are in English. Not only will students want to be credentialed in English, they will want these credentials from an English-speaking country. So international student numbers will certainly increase, particularly in English-speaking countries.

Another trend though might be that other countries will accept or require theses done in English, or another "international" language. For example, Denmark is a non-English-speaking country where one can write one's thesis in English. The research culture is also spreading, so mathematics education will grow in its research importance in those universities. There will still be some cultural differences, but there are no signs on the horizon that doctoral studies in mathematics education will decline in any way.

CAN WE CREATE MORE INTERCULTURAL COLLABORATION IN DOCTORAL PROGRAMS?

Note first of all that I have changed the word *international* to *intercultural*. Whilst it does seem sensible to focus on certain differences between, and relations between, countries, I certainly believe that it is more interesting and profitable from a research perspective to consider intercultural collaboration. The growing diaspora situation is also making us more aware of cultural differences and similarities.

For a start, we know a little about cultural conflicts and culture shock experienced by international students, and we don't call it nation shock! It is the cultural differences that are so important and revealing: language, customs, history, values, etc. Furthermore, I would argue that we must create more intercultural collaboration if we wish to avoid the kind of cultural imperialism that has plagued "Western" mathematics itself.

We have to debate the issues of diversity and comparability since every field, particularly a predominantly practice-oriented field like ours, needs multiple perspectives in its research approaches. At the same time, we are talking here about research students, and to be fair to them we need to be sure about comparability and standards. There are certainly differences between countries in terms of their

- entry requirements (e.g., language, exam performance, educational experiences)
- teaching and supervisory styles,
- examination procedures.

Also we need to pay more attention, not less, to the cultural and social perspectives of our international research students. By this means we are not only better able to teach and encourage them, but also we, and our other students, can learn from them and so enrich our own cultural knowledge and experience.

We need to encourage more collaboration between research students across the cultural divides, through student exchanges, distance-learning activities, paper exchanges, and so on. The World Wide Web and e-mail allow for much more communication and collaboration across the world and should be exploited by us much more. After all, if being a research student is an acculturation experience, then we need to ask, "Into what culture are they being inducted?" It is surely the international culture of mathematics education, and that is largely accessible these days through the internet.

We should also be striving to help colleagues in other countries and cultures to get their own doctoral programs started. This requires much tact and diplomacy, particularly as staff in those universities may well have received their doctorates from overseas universities. Also we need to be ready with counter arguments to our own administrators who may suggest that this development could reduce the numbers of international students coming to our own universities!

The issue eventually boils down to whether one believes that cultural diversity in research is an asset or an obstacle to progress. I have absolutely no doubt in my own mind, and I would hope that those attending this conference would also have no doubt. After all, I assume that we are all here because of our shared belief that diversity is beneficial to knowledge development. Cultural, and national, diversity is for me clearly part of that belief, and a very strong part.

SOME USEFUL INTERNATIONALLY ORIENTED WEBSITES FOLLOW:

http://www.stolaf.edu/other/extend/
    Resources/resources
http://www.nottingham.ac.uk/csme
http://elib.zib.de
http://www.emis.de/ZMATH.html
http://www.fi.ruu.nl
http://ued.uniandes.edu.co
http://camel.math.ca/Education
http://www.bp.com/saw/english/core.html
http://sunrise.eng.monash.edu.au/MERGA
http://www.aamt.edu.au

Alan J. Bishop
Faculty of Education
P.O. Box 6
Monash University
Victoria 3800, Australia
alan.bishop@education.monash.edu.au

PART 2: CORE COMPONENTS

CBMS Issues in Mathematics Education
Volume 9, 2001

DOCTORAL PROGRAMS IN MATHEMATICS EDUCATION: FEATURES, OPTIONS, AND CHALLENGES

James T. Fey, University of Maryland

Doctoral study in mathematics education has long been identified as the professional preparation for careers in higher education that emphasize teacher education and research on mathematics teaching and learning. However, over the past several decades we've seen the emergence of many new professional positions in mathematics education that really require the kind of advanced education provided by doctoral studies. This broadening of the purposes of doctoral studies makes it timely to examine the structure of university graduate programs in mathematics education.

WHAT ARE THE ISSUES?

Let me start by offering a framework of program dimensions and then elaborate each with some personal concerns. It seems to me that design and operation of a doctoral program has to consider questions in five broad areas:

- What is the purpose of doctoral programs in mathematics education? What are the professional roles that doctoral students are preparing for?
- What knowledge, abilities, and dispositions do those students need to acquire?
- How can graduate programs in mathematics education provide the essential knowledge and personal development?
- How can graduate programs assess the competence of their doctoral candidates?
- How can students be recruited to appropriate programs and placed in appropriate professional positions when graduated?

PROFESSIONAL GOALS OF DOCTORAL STUDENTS

The natural first question to ask is why have doctoral programs in mathematics education at all? What services do such programs provide, and who are their natural clients?

In most university academic fields (like mathematics, history, physics, or English) the doctoral degree is *assumed to be* preparation for a career in higher education that would combine teaching and research. Typical doctoral programs in those areas provide only modest development of their students' teaching skills, while focusing attention on preparation for research. We all know that this research-oriented Ph.D. model doesn't really match the ultimate careers of most graduates, but the situation in a professional field like mathematics education is also different from the start.

Some who earn the Ph.D. or Ed.D. in mathematics education follow career paths that are very similar to those of doctoral graduates in academic fields. They teach in colleges or universities, and they do research on basic questions of learning or teaching. However, there are many others who apply their advanced study in mathematics education to careers that emphasize practical teacher education, leadership in school systems, curriculum and assessment development (including technology applications to education), program evaluation, mathematics teaching in two-year or four-year colleges, and policy leadership in national government or professional associations. The base of operations for such professional mathematics educators might be a college or university, an independent research and development organization, a government agency, a local school system, an education-related business, a foundation or nonprofit organization, or an independent consulting firm.

It seems to me that one can make a very strong argument for doctoral-level study in mathematics education as preparation for any of the careers I've mentioned. For example, a very interesting new doctoral program at Montclair State University expands conceptions of doctoral study to include professional mathematics educators who remain in the classroom, providing high-level leadership at the school level.

KNOWLEDGE, SKILL, AND DISPOSITIONS

Regardless of our particular professional positions or bases of operation, most of us in mathematics education are expected to have a broad range of knowledge and skills that allow us to connect and contribute to the various communities with interest in and expertise in mathematics. Most of us are idea brokers whose work takes place in several quite different arenas. Given the variety of professional positions in which doctoral graduates in mathematics education might land, it's not easy to define a list of knowledge and skills that everyone would agree is essential. As I have reflected on my own career and what I see others in the field doing, I can see a strong case for the following elements:

- Breadth and depth of knowledge in mathematics and its applications at the level we expect to focus our professional work and several years before and beyond. This knowledge of mathematics should include perspectives that go beyond technical skill- awareness of mathematical habits of mind, something of the historical development of big ideas and the contributions of different cultures, personal confidence in one's ability to do and learn mathematics, and a disposition to keep growing mathematically.

- Understanding of how people learn various aspects of mathematics at various stages of development and a disposition to keep growing in this understanding.

- Skill and experience as a teacher of mathematics with students at the levels of education that one's professional role focuses on and a disposition to expand one's repertoire. This instructional repertoire should probably include use of the emerging technologies for teaching and learning.

- Understanding of the broader educational and social context for mathematics education, including main themes in its history.

- Skill in working with teacher candidates, in-service teachers, and schools in professional development and leadership activities. This skill is enhanced if one understands the dynamics of change by individuals and organizations and if one is able to communicate ideas clearly and persuasively.

- Knowledge of the research basis for practices in mathematics education.
- Scholarly skills needed to contribute to improvement of mathematics education through research, interpretation of research for practice, curriculum development, and policy analysis.

I suspect that for many readers, my suggested domains of knowledge, skill, and disposition describe a daunting wish list that few doctoral graduates master. I wouldn't begin to argue that the doctorate requires mastery of every item on the list of desirable knowledge and skills. The tricky problem is defining what is enough for a beginning professional at the doctoral level (because I'm quite sure that we aren't interested in having the doctorate defined as a kind of culminating accomplishment in one's field).

At the same time I've defined a very broad and demanding list of knowledge and skills for mathematics education doctorates, I suspect that many will feel that the list does not give proper attention to the centrality of research in doctoral study. Let me state my personal point of view in defense of a more eclectic list of priorities.

There is a historical tradition that the doctoral degree recognizes some outstanding contribution to human knowledge, and there are striking examples in mathematics of such groundbreaking dissertations. It's much harder to identify such contributions by doctoral students in mathematics education or, for that matter, in any discipline during the past half century. Much as we mathematics educators (or educators in general) would like respect as legitimate members of our university communities, with recognition of the worth in our scholarship, I believe that a professional field like ours is different from traditional academic disciplines. It requires different talents and different kinds of knowledge, and it makes different kinds of contributions to society. Ours is an *appropriately eclectic field*. It requires advanced study and skills comparable to those in other professional fields that award doctorates, but a narrow focus on research isn't the right thing.

In the professional work that I've done recently, my responsibilities have included almost every one of the roles on the earlier list-organizing curriculum development, guiding workshops for teachers, forming policy guidelines for teacher education in mathematics, doing empirical research, teaching mathematics, teaching future teachers, participating in public meetings with school communities, writing persuasive essays for news media and practitioner journals, writing historical analysis of issues in our field, and so on. One could say that I am arrogant to think that I can contribute in so many arenas. But I suspect that most active mathematics educators have done the same things.

STRUCTURE AND OPERATION OF DOCTORAL PROGRAMS

If the goals of doctoral programs in mathematics education are different from those of traditional academic disciplines, the practical constraints under which mathematics education programs operate are also different from those of doctoral programs in traditional academic fields. Typical mathematics education doctoral programs have only a few entering students each year. Most mathematics education programs have only five or six faculty, whose energies are drained by heavy teaching responsibilities in teacher preparation programs. In even the largest mathematics education doctoral programs, many students work full-time and study part-time. Furthermore, the few doctoral students that most programs enroll often have quite diverse goals for their doctoral studies.

Given the tendency of American universities to allocate resources according to weighted-credit-hour formulas, it is very difficult to provide appropriate graduate courses in mathematics education at even several dozen universities. *It seems to me that one of the most challenging problems facing doctoral programs in mathematics education is offering high-quality education with constrained faculty expertise and small numbers of students.*

How do we provide mathematics education doctoral students with advanced content knowledge and perspectives, when typical graduate mathematics courses aim at the kind of specialized technical knowledge useful primarily for dissertation research in mathematics? It would be very nice to have doctoral students participate in seminars that examine school mathematics from an advanced standpoint, but how many of us have the right mathematics faculty or the numbers of eligible students to operate such programs? I'm sure that I learned much more about school mathematics from participation in a series of curriculum development projects than from graduate analysis or statistics courses at Columbia University. But how do we make those kinds of experiences widely available?

How do we provide students with in-depth understanding of the history and current problems of curriculum, teaching, and research in mathematics education? We are fortunate to have a rich literature on these issues, with more forthcoming; so students can get much of this knowledge through independent reading or tutorial arrangements with faculty. However, few universities have faculty with expertise in all aspects of this background knowledge.

How do we provide students with theoretical and practical knowledge of important research methods? How do we help them develop skills in teacher education and professional development of in-service teachers? Some of this knowledge and practical skill can be acquired in the generic courses that most research universities offer in their colleges of education. But, once again, I suspect that most of us view participation in real research projects, internships in teacher education courses, and teamwork in workshops for teachers as the most effective ways to develop the skills and dispositions required by subsequent research and teacher education responsibilities.

How do we help doctoral students become capable and active scholars who contribute to local, regional, and national research and educational policy communities? Our field is not unlike most academic fields in the sense that few doctoral graduates become consistently productive scholars. In part that is a consequence of the professional responsibilities that they face in their postgraduate positions. But it also reflects an unfortunate fact of life about the gap between "school scholarship" and scholarship in the real world.

The challenges of making doctoral education in mathematics education an authentic and effective preparation for productive professional careers call for new ideas of how doctoral programs are conceived and organized. Of course, one possible solution to the "critical mass problem" would be to recommend that only a few universities should sponsor doctoral programs in a specialty like mathematics education. I'm afraid that for most of us the politics of higher education make this an unattractive option; it would also have the effect of denying doctoral study to many capable students who couldn't manage the logistics of a cross-country move and the full-time study that would probably be involved.

To me this situation cries out for interinstitutional collaboration, making the best of each university's faculty and programmatic resources available beyond its usual enrollment boundaries. We probably need to think less about conventional courses and more about active engagement of doctoral candidates in the wider professional community, but that solution is easier to propose than to accomplish.

As I've reflected on the educational experiences that have meant the most to me and to graduate students I've worked with, there is an array of activities that can be very productive.

- Courses can be effective for acquisition of many kinds of knowledge. To provide the best that our field has to offer in an economically viable format, it seems possible to form consortia whose faculty would share responsibility for offering key courses through distance learning formats, short courses, or special institutes.

- Field experiences and internships with professional development and curriculum development projects are really the most effective ways to develop skills in those fundamental activities that doctoral graduates will be called on to perform. Again, it makes sense for consortia to collectively arrange those sorts of opportunities for their students.

- In today's educational environment there are many professional mathematics educators active in policy-making roles in state and national governments and in national professional organizations. Again, it makes sense to have a consortium arrangement that would provide internship opportunities in those settings for interested doctoral students.

- Many academic disciplines do a very minimal job of preparing their doctoral students for their primary postgraduate work-teaching. It seems to me that for doctoral students in mathematics education, such preparation for future teaching responsibilities is especially important because our courses (and workshop leadership responsibilities) are much more difficult than typical content courses. Internship experiences, working alongside senior faculty and experienced school-based leaders, are powerful learning opportunities, worth far more than any formal coursework.

- While the typical dissertation is a solo, one-time research ordeal, doctoral students who learn research skills through participation in substantial projects led by senior faculty are often those who remain active in scholarship after completion of their graduate studies. It seems quite possible for our professional organizations to facilitate connection of doctoral students and on-going research projects-to the benefit of the research projects and the students.

I'm quite sure that for most readers of the preceding consortium proposals, first thoughts will include dozens of reasons why they can't work. Turf and logistics are two of the immediate grounds for concern. However, if we view consortium arrangements as opportunities to provide all students with the highest level expertise available on different campuses, it is likely that each doctoral university faculty will have things to contribute. Furthermore, the professional interactions induced by planning for and delivering consortium-based programs would be stimulating for all of us.

ASSESSING QUALIFICATIONS OF DOCTORAL STUDENTS

At our university, two of the criteria by which graduate programs are rated are the GRE scores and GPAs of *entering* students. It's not completely clear to me that these numbers are the best indicators of success in graduate school or in the future professional careers of mathematics educators, but measurements of exit accomplishments do seem important and relevant. Traditional criteria for completion of doctoral programs include oral and/or written qualifying examinations and successful defense of an independent dissertation. Both sorts of examinations are being challenged-I think with good reason. My concerns about traditional written examinations and the dissertation are captured in a simple question: *Are they authentic and predictive measures of the competence that we want graduates of our programs to have?*

I'll have to admit to some persistent ambivalence about the purpose and format of traditional written qualifying or comprehensive examinations. On one hand it seems silly to require students to take courses covering what we judge to be the core material and then ask them to take what amounts to a second final exam on the same material. Furthermore, we can't begin to examine students on everything we hope they know about mathematics, mathematics education, and research, and the criterion for satisfactory work has shaky validity.

On the other hand, I really do believe that there are things about our field that anyone earning a doctorate ought to know, and I believe that a doctoral student should be able to organize ideas into coherent presentations. I hope that the comprehensive examination questions I ask our students at Maryland require thoughtful integration of knowledge and identification of big ideas in the field. Nonetheless, even the best written examination questions don't adequately assess many aspects of a professional mathematics educator's essential repertoire of skills and dispositions.

With a relatively small number of doctoral students and a relatively small faculty, the two groups almost always get to know each other pretty well. At Maryland we usually see to it that our doctoral students work alongside the faculty in teaching courses and in the research, curriculum development, and teacher professional development projects we are involved in. Reading lists and written examinations seem a poor alternative to such genuine involvement in professional work. But it might make a very stimulating contribution to the field if some consortium of faculty made regular recommendations on the broad outlines of information that doctoral students in mathematics education should be expected to master.

The assessment issue that generated the most discussion at this meeting was the doctoral dissertation. The challenge of identifying a suitable problem, gaining access to appropriate research sites and subjects, writing the typical long document, and defending the report successfully is typically a greater barrier to completion of the doctorate than any course or examination. Furthermore, many students who do manage to complete the dissertation develop such distaste for the whole process that they never embark on another research study.

Over the past decade I've been thinking about a variety of alternatives to the traditional "great work" conception of a dissertation. The recent article by Nell Duke and Sarah Beck (1999) articulated many of the ideas that I've been talking about with colleagues. Duke and Beck suggest two criteria for judging the value of any particular dissertation format.

- Will the chosen format make it possible to disseminate the work to a wide audience?
- Will the project help prepare candidates for the type of scholarship they will be expected to do throughout their careers?

As alternatives to the traditional dissertation project and long five-chapter report, they expand on ideas from Krathwohl (1994) to propose that a dissertation might better consist of several articles ready for publication. The articles might be related research reports on aspects of one line of investigation. They might include a combination of research articles and interpretations of the results suitable for practitioner journals. They might be a collection of policy memos intended for governmental or school officials.

Duke and Beck are careful to caution against losing virtues of the traditional dissertation format. But in my view the risks of such a move are far outweighed by the opportunity to make that aspect of doctoral study a much more authentic preparation for future scholarly work in education. Duke and Beck note that education is not the only field considering alternatives to dissertation traditions. In fact, they cite a number of fields in which collections of published work have long been acceptable formats for dissertations. The whole idea of alternatives to the traditional dissertation seems an enormously attractive idea.

CONNECTING DOCTORAL STUDENTS TO PROGRAMS AND THEN JOBS

Several of my proposals concerning delivery of doctoral program courses and experiences recommended interinstitutional collaborations that are not very common in our field today. One obvious reason to be skeptical of prospects for such proposals is the natural American habit of competing rather than cooperating. We compete for students, we compete for faculty, we compete for research funds, and we compete for reputation. Some of that competitive posture is forced on us by external factors like rating games and pressures from our universities to collect students and money, but some of it comes from the culture of our professional communities by which we are too easily seduced.

I'm not sure that I have any bright ideas for how we can help prospective doctoral students identify the university graduate programs that will be right for them. None of us have large enough programs that we can afford extensive advertising campaigns for students. I suspect that word-of-mouth and geographic proximity are the primary factors in graduate school decisions by most students. It would be nice if there was some central web-site to help potential students in their first pass at identifying graduate programs that match their goals and experiences and aptitudes.

What would also be helpful in our field is a more rational system for connecting doctoral graduates to appropriate professional jobs. For those who seek college or university mathematics education positions, determining the availability of such opportunities requires skills and persistence unlike those required by the jobs themselves. For those who seek to hire a new faculty member, the search proceeds with very limited information about the candidate pool and the timing of other competitive searches. I suspect that this disorderly process often sabotages institutions' search efforts, despite great investments of time and energy.

Finally, once new doctorates are connected to faculty positions, the scramble for tenure and promotion begins. Here I think the mathematics education profession probably does a reasonably good job of providing outlets for scholarly work and meetings to help new scholars become part of an active research community. The research pre-session at annual NCTM meetings provides regular opportunities for researchers, including recent doctorates in mathematics education, to share and exchange ideas. At the same time, it seems quite reasonable for our community to extend further in providing opportunities for collaborative work and publication that draw the new doctorates into careers of productive work.

CONCLUSIONS

The phrase "challenge and opportunity" is often used to describe problematic situations that are more challenge than opportunity. However, it seems a very appropriate way to describe conditions in mathematics education doctoral programs today. Mathematics continues to be one of the most highly valued strands of the school curriculum. This high regard usually translates into a demand for well-educated professionals who can assume significant academic, school, and governmental positions. There is a strong positive sense of community among current mathematics educators, and we have the communication tools to apply the energy and expertise of that community in creative ways to provide attractive and effective new doctoral programs. Productive innovation will require some thinking and acting in new ways. But, it seems to me that there is enormous potential payoff for those of us already in the field and for the doctoral students who will become our colleagues and successors.

James T. Fey
Department of Mathematics
University of Maryland
College Park, MD 20742
jf7@umail.umd.edu

CBMS Issues in Mathematics Education
Volume 9, 2001

THE RESEARCH PREPARATION OF DOCTORAL STUDENTS IN MATHEMATICS EDUCATION

Frank K. Lester, Jr., Indiana University
Thomas P. Carpenter, University of Wisconsin

At a conference held at Cornell University in 1968 and attended by a small group of prominent research mathematicians and mathematics educators, participants agreed that "for Mathematics Education to consolidate its place as a respected scholarly field, both the research and the people involved must necessarily be of high quality. Hence, one faces the problem of providing graduate training for its potential practitioners" (Long, Meltzer, & Hilton, 1971. pp. 451-452). It seems that serious discussion of the nature of the research preparation received by mathematics education doctoral students began more than 30 years ago. Since that time, many doctoral programs in mathematics education have been developed—some clearly designed to prepare serious researchers, others with no such intention.

Our aim in this short paper is to summarize the key ingredients of the focus group discussions about the research preparation of doctoral students held at the National Conference on Doctoral Programs in Mathematics Education.

CONFERENCE DISCUSSIONS OF RESEARCH PREPARATION

The discussions were organized around five topics: areas of knowledge, research-related coursework, research traditions and methods, the role of the dissertation, and the development of research communities.

AREAS OF KNOWLEDGE

Participants agreed that the research component of a doctoral program in mathematics education could be thought of as focusing on the development of knowledge in three areas: (a) "core" knowledge for all students, (b) knowledge in a student's own area of specialization, and (c) knowledge related to the specific topic/questions of a student's dissertation. These three areas of knowledge can be visualized as a set of nested rings (Figure 1). Each student is expected to acquire the core

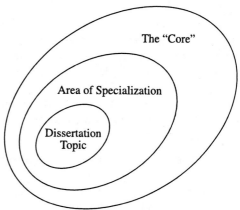

Figure 1. Areas of research knowledge for doctoral students

knowledge (the outer ring) through coursework, apprenticeships, and independent study. In addition, each student is expected to identify an area of special interest about which he or she will become especially knowledgeable (the middle ring). Finally, within the particular area of specialization, the student will focus on a topic/problem that he or she will investigate deeply via the dissertation (the innermost ring).

Discussions of these areas of knowledge led to consideration of three fundamental questions:

- Can a research "core" be identified? If so, what is it, and how should it be learned?
- How should knowledge in the middle ring be acquired?
- How should students be assisted in their pursuit of knowledge in the innermost ring?

As expected, participants were unable to reach agreement on either a core or the means for enabling students to acquire knowledge of any of the three areas. A reasonable reading list that all entering doctoral students could be given would include the *Handbook of Research on Mathematics Teaching and Learning* (Grouws, 1992), supplemented by a small collection of more recent research reviews and carefully selected classic research reports. Such a set of research syntheses and reports would well equip students with basic knowledge about research emphases, methods, trends, and results.

COURSEWORK & OTHER RESEARCH-RELATED EXPERIENCES

In a 1971 article in the *American Mathematical Monthly*, Walbesser and Eisenberg argued against the practice in many (perhaps most) programs of restricting doctoral students' research training to "courses in applied statistics, educational measurement, experimental design, and summaries of various learning theories" (p. 668). Although current programs do not appear to restrict students' research training to courses such as those identified by Walbesser and Eisenberg, it seems that some (perhaps too many) programs require their students to take a smattering of generic quantitative and qualitative courses rather than a thoughtfully considered set of courses that helps them learn about the full range of activities in which active researchers engage.

In the same *Monthly* article, Walbesser and Eisenberg insisted that "vicarious encounters with research are not viable substitutes for direct involvement... If one is to become a researcher, then his research involvement must be at the participating level rather than the spectator level" (p. 669). Conference participants generally agreed with Walbesser and Eisenberg's comment, but they did not agree on how to get students directly involved in research activities prior to their dissertation. In fact, apart from coursework focusing on research, there was no consensus as to the sorts of research experiences doctoral students should receive.

By contrast with the situation 30 years ago, contemporary programs aim to prepare students who not only are well grounded in one research tradition (i.e., capable of conducting research in that tradition in a knowledgeable manner) but also are familiar with at least one other tradition (i.e., able to interpret research conducted in that tradition, but not necessarily qualified to undertake such research). With this expectation in mind, programs often require students of today to take a beginning course dealing with strategies for conducting educational research, fundamentals of statistics applied to educational research, fundamentals of qualitative inquiry in education, a second course in either quantitative or qualitative methods, and a seminar on reviewing research literature and writing a research proposal. On occasion these courses complement various kinds of apprenticeship experiences (see discussion of apprenticeship training below).

RESEARCH TRADITIONS AND RESEARCH METHODS

Because mathematics education borrows liberally from several disciplines, it is essential that the preparation graduate students receive includes attention to the research traditions of several disciplines (anthropology, psychology, and sociology are prominent examples). But, is it reasonable to expect any doctoral program in mathematics education to provide adequate preparation in conducting research based on so many different traditions? Although time limited the opportunity to consider this question at length, participants acknowledged it as one deserving extensive deliberation.

ROLE OF THE DISSERTATION

According to Lester and Lambdin (in press), before about 1980 the dissertation was begun at the end of a student's coursework and often was viewed as the final hurdle to be cleared before earning an academic "union card." Today, those doctoral programs that emphasize the preparation of researchers regard the dissertation as a culminating research activity following three or more years of research-related experiences. In general, however, the role and importance of the dissertation in the research preparation of doctoral students vary from institution to institution. At some institutions the dissertation is the sole research experience in which students engage.

DEVELOPING COMMUNITIES OF RESEARCHERS

In a paper prepared for the National Academy of Education's Commission for Improving Education Research, Alan Schoenfeld (1999) suggests that "students are more likely to become productive researchers, and to develop useful habits and perspectives more rapidly, if they are members of a research *community*" (emphasis in original). As reasonable as Schoenfeld's comment may be, it also poses a dilemma for mathematics educators who are responsible for doctoral programs but who are not engaged in research themselves. It seems clear that students who have worked alongside experienced researchers are much more likely than those who have not to become active researchers.

In order to provide students with the sort of enculturation necessary to be active participants in a research community, Schoenfeld insists, and we agree, that students should serve apprenticeships in a research community from the beginning of their program of study. When properly applied, this recommendation would mean that students not only would develop "useful habits and perspectives more rapidly," but would also become familiar with the special language associated with a particular research tradition and acquire an awareness of the underlying assumptions and purposes of research conducted within that tradition. Unfortunately, apprenticeship training is relatively rare in mathematics education doctoral programs today. In their study of mathematics education doctoral programs, Lester and Lambdin (in press) found that until quite recently (the 1990s) doctoral students typically have not served any kind of real research apprenticeships and, consequently, have had little or no opportunity to develop a sense of any legitimate research tradition. Indeed, the practice of restricting students' pre-dissertation research experience to various discrete courses remains the norm in most mathematics education doctoral programs.

CLOSING COMMENTS

It is apparent to us that the nature of mathematics education doctoral study varies widely across the United States; in particular, the nature of the research preparation doctoral students receive differs dramatically from university to university. At some—perhaps many—institutions, the only direct involvement in research required of doctoral students is the dissertation. At others—perhaps only a few—doctoral students grapple with research problems and issues throughout their programs of studies, and the dissertation stage serves as an indicator of the students' readiness to enter the world of research on their own.

The growing complexity of mathematics education research demands a corresponding change in the preparation of those who would engage in this activity. In our view, doctoral programs that choose to emphasize apprenticeship training in addition to formal coursework are most likely to develop future generations of mathematics educators who will be able to conduct successful, productive research. But, because the job market for highly qualified mathematics education researchers will almost assuredly be quite limited for the foreseeable future we suspect that substantive apprenticeship training will remain common at only a small number of universities.

Frank K. Lester, Jr.
Education 3056
Indiana University
Bloomington IN 47405
lester@indiana.edu

Thomas P. Carpenter
University of Wisconsin
Department of Curriculum and Instruction
225 N. Mills St.
Madison WI 53706
tpcarpen@facstaff.wisc.edu

CBMS Issues in Mathematics Education
Volume 9, 2001

THE MATHEMATICAL EDUCATION OF MATHEMATICS EDUCATORS IN DOCTORAL PROGRAMS IN MATHEMATICS EDUCATION

John A. Dossey, Illinois State University
Glenda Lappan, Michigan State University

The mathematical education of students in doctoral programs in mathematics education is part of the tripartite world of teachers' knowledge that Shulman (1986) so artfully described. It constitutes the content-specific knowledge that these prospective mathematics educators ought to have. There is no simple listing of courses that will describe this aspect of doctoral programs in mathematics education. Rather, one has to examine the question from the standpoint of "what mathematics" for "whom." Mathematics education programs prepare individuals for a wide variety of professional careers. These roles include researcher in a university setting, curriculum director for a school system, teacher educator at a state university, mathematics/mathematics education professor in a smaller college, and mathematics specialist with a textbook or measurement company. The requirements and roles associated with these career options, and others, vary. What is important is that the individual student be able to "do mathematics" at an appropriate level.

What does it mean to be able to "do mathematics" at an appropriate level for a doctoral student in mathematics education? It means that the student is able to:

- appreciate the rules of evidence within the discipline;
- outline and connect the major ideas of areas within the discipline;
- analyze and apply the major algorithms and procedures within the discipline;
- describe the ways of thinking through which the discipline itself expands;
- use disciplinary knowledge to solve problems in the discipline and related disciplines; and
- see the connections among and between ideas, concepts, structures, and methods within the discipline and outside the discipline.

In addition, particular knowledge that is beyond, but connected to, these ideas is needed by individuals headed in certain directions. For example, individuals preparing for futures in teacher preparation will need deep knowledge of school mathematics and its relationship to mathematics that they study at higher levels.

Setting the levels of disciplinary knowledge to assure that graduates of doctoral programs are capable of meeting these criteria requires great care. In the past, many doctoral programs required the equivalent of an M.S.+ 30 semester hours in pure mathematics and a written examination of all students. At the same time, other programs had essentially no specific requirements concerning mathematics outside of the course work in professional mathematics education-the content-specific knowledge area of doctoral programs in mathematics education.

To begin to narrow the question of what knowledge of mathematics is needed, expressing some assumptions about the nature of the requirements for all doctoral students might be an appropriate avenue. Klein's (1932) notion of knowing elementary mathematics from an advanced standpoint comes to mind as a major criterion of what students ought to know. Other criteria include mathematics that forms a basis for:

- guiding students in exploring mathematical situations in a constructive manner, determining which involve meaningful mathematics and which do not;
- analyzing school mathematics and the nature of issues in and around changes in its content and learning at a depth that allows for observing and working with students and creating curricular materials; and
- applying the content of mathematics (concepts, principles, procedures, problem-solving strategies, reasoning, communication modes, representations, etc.) to problem situations commensurate with a level that is at least six educational grade-levels above that of their teaching assignment.

The last criterion suggests that a mathematics educator preparing to teach prospective elementary teachers ought to have a command of mathematics at a level equivalent to that of the first three years of undergraduate work in mathematics (Grade 9 + 6). This criterion assures that these elementary-focused mathematics educators have a mathematics background sufficient to engage in conversation with teachers over the full range of K–12.

The criterion suggest that teachers of prospective secondary school teachers ought to have a grasp of mathematics through an M.S.+ 30 in mathematics (Grade 12 + 6). However, the courses and programs that optimally lead to meeting these criteria might differ significantly from those currently in place in many institutions offering doctoral programs in mathematics education. This criterion of "plus six" means that graduates can discuss articulation issues across the boundaries of elementary, secondary, and postsecondary education in meaningful ways.

SAMPLE PROGRAMS OF MATHEMATICS FOR DOCTORAL CANDIDATES

The following sections outline what might be considered solid preparations in mathematics for candidates in mathematics education. They are outlined in terms of courses with notes on the content and delivery modes for such courses. Meeting the needs of mathematics educators will require a transformation of the undergraduate and graduate programs for such individuals. However, the transformations described are in line with current recommendations for changes in mathematics and science education at the collegiate level (Committee on Undergraduate Science Education, 1999).

Mathematics department faculty must be brought into the discussion of the transformation of courses to meet the needs of mathematics majors, prospective teachers of mathematics at every level, and prospective mathematics educators. If we are to break the chain of learning and teaching as one learned and was taught, we must restructure courses for mathematics educators, future teachers, and other undergraduates in mathematics. Such a refocusing must result in a more constructive classroom, one that is centered on important mathematics and that meets the primary goal of making sense of and connecting mathematical ideas.

MATHEMATICS FOR DOCTORAL STUDENTS SPECIALIZING IN TOPICS RELATED TO ELEMENTARY SCHOOL MATHEMATICS

Doctoral students preparing for situations where they will focus on topics found in elementary school mathematics should have an understanding of mathematics equivalent to that of an individual getting a strong undergraduate minor in mathematics. However, the course work in such a program must go beyond simply completing the current course work related to such a level of understanding.

The creation of new, carefully constructed programs of study is required to serve this need. Such courses need to focus on the concepts, principles, and procedures associated with the following courselike grouping of topics. Such course work should focus not only on the development of grasp of the content, but also on the connections and manner in which the content itself is developed from first principles. The courses should reflect the spirit of the original discovery of the material and the current applications of the material in mathematics and related disciplines. Students should have multiple opportunities to experience developing and extending the mathematical ideas. Such courses should reflect a degree of depth, breadth, and sophistication not seen in the traditional undergraduate courses for majors.

For individuals preparing for doctoral studies in mathematics education specializing in elementary school mathematics issues, the course work would focus on the topics usually covered in courses in:

- differential and integral calculus with a focus on the meaning and applications of change and accumulation;
- abstract algebra with a focus on the ring of integers, isomorphism, field of real numbers, and polynomials over the real numbers;
- linear algebra with a focus on systems of equations, independence, and matrix transformations;
- elementary number theory through linear congruence;
- discrete mathematics with an emphasis on counting, finite probability (including simulations), and graph theory;
- geometry from an investigative viewpoint; reviewing Euclidean geometry from an advanced standpoint and the development of an deep understanding of transformations/tessellations;
- statistics with an emphasis on exploratory data analysis, simulations, and univariate and bivariate distributions; and
- history of mathematics with an emphasis on development of number, geometry, and algebra.

Here is a good place for course designers to try to flesh out the structure of new courses that look at depth of understanding of elementary topics in the K–12 curriculum.

MATHEMATICS FOR DOCTORAL STUDENTS SPECIALIZING IN TOPICS RELATED TO MIDDLE SCHOOL MATHEMATICS

Doctoral students preparing for situations where they will focus on topics and issues related to middle school mathematics should have a background in mathematics equivalent to that of an individual getting a strong undergraduate major in mathematics. Their course work should draw on a combination of the regular courses offered for mathematics majors and specialty courses, where possible, for the doctoral students specializing in elementary school mathematics topics and issues listed above. These doctoral students' programs in mathematics should provide the content they will need for their research and teaching and prepare them for continued study of mathematics as a discipline.

As a result of the growing emphases in algebra and geometry at the graduate level, doctoral students specializing in middle school mathematics should complete a set of courses equivalent to:

from the mathematics offerings–

- differential and integral calculus taught from a reform calculus standpoint;
- linear algebra;
- introduction to number theory;
- discrete mathematics with an emphasis on counting, finite probability (including simulations), and graph theory; and
- history of mathematics with an emphasis on development of number, geometry, and algebra.

from offerings designed for students in mathematics education–

- a semester of abstract algebra with a special focus on the number theoretic aspects of the integers, the development of rings, integral domains and fields (including the real and complex numbers), and polynomials over the real numbers;
- geometry from an investigative viewpoint, reviewing Euclidean geometry from an advanced standpoint, developing finite and non-Euclidean geometries, development of deep understanding of the role of geometry from a transformational approach; and
- statistics with an emphasis on exploratory data analysis, univariate and bivariate distributions, hypothesis testing, and regression.

Doctoral students preparing for this level should develop a deep understanding of both algebra and geometry. In so doing, they should have additional course work in algebra and analysis that will provide them with a solid understanding of the function concept and the role of variable in expressing growth and change. Course work in technology and its application in modeling mathematical situations would be a solid alternative for such students.

One example of a special course that might be designed for graduate students in mathematics education at this level is a course in combinatorics created by Roger Day (1999).

MATHEMATICS FOR DOCTORAL STUDENTS SPECIALIZING IN TOPICS RELATED TO SECONDARY SCHOOL MATHEMATICS

Doctoral students preparing for situations where they will focus on topics and issues at the level of secondary school mathematics should have, at minimum, an M.S. in mathematics. Their course work should have not only the same broad coverage of topics as those listed for previous specialty levels, but also additional work in number theory, algebra, analysis, geometry, and probability and statistics. Such students should have a solid understanding of the elementary number theory through quadratic reciprocity and contemporary applications of number theory; a solid understanding of the fields of real and complex numbers and polynomials over them through the Fundamental Theorem of Algebra and Galois theory. In addition, they need a solid understanding of analysis of real variables and an introduction to complex variables, as well as work in differential equations and extended course work in geometry and probability and statistics. These students' backgrounds should include additional work in the history of mathematics through the twentieth century and some course work in modern applied mathematics (linear programming, algebraic coding theory, discrete dynamical systems, mathematics of finance, etc.).

MATHEMATICS FOR DOCTORAL STUDENTS SPECIALIZING IN ADVANCED MATHEMATICAL TOPICS

Doctoral students preparing for situations where they will focus on topics and issues at the undergraduate level or beyond should have at least an M.S.+ 30 in mathematics or beyond. This preparation should have a solid, broad M.S. plus the first full year of graduate study in algebra, analysis, probability and statistics (including some nonparametric statistics), two applied areas, and additional discrete methods.

SUMMARY

The foregoing recommendations were made with the realization that each doctoral program exists in a special milieu—one that depends on intradepartmental cooperation and trust. Faculty involved in doctoral programs in mathematics education and in mathematics must work together to establish sequences of course work that do not replicate the status quo. These programs, and the courses in them, must provide both new models for deep understanding of the content and its applications and models of how that content can be acquired in a constructive learning environment.

This is not a call for weakening the mathematics requirement. In reality, it is a call to develop individuals in mathematics education who will know mathematics deeply, who will be focused on content in the areas in which they have the responsibility to prepare or support teachers, and who will meet the original goals and criteria mentioned at the beginning of this paper.

The task is to design new sequences of courses for doctoral students in mathematics education. Such courses should be developed to provide flexibility within programs sufficiently large to provide the differentiation described above. When a doctoral program is unable to provide the specialties discussed, efforts should be made to see that everyone reaches the level of the M.S. in mathematics. However, for large programs in which specialization is possible, efforts should be made to tailor programs of study in mathematics to maximize the focus on important mathematics that will optimally prepare doctoral students for their later work in research, teaching, and service to the profession. This is an important task and one that will require as much dedication and effort as the creation of the course work in methods courses, core courses, and research courses in the doctoral programs in mathematics education.

John Dossey
Illinois State University
R.R. #1, Box 165
Eureka, IL 61530
jdossey@math.ilstu.edu

Glenda Lappan
A-718 Wells Hall
Michigan State University
East Lansing, MI 48824
glappan@math.msu.edu

CBMS Issues in Mathematics Education
Volume 9, 2001

PREPARATION IN MATHEMATICS EDUCATION: IS THERE A BASIC CORE FOR EVERYONE?

Norma C. Presmeg, Illinois State University [1]
Sigrid Wagner, Ohio State University

As far as course work is concerned, the possibilities for doctoral programs in mathematics education are wide-ranging. These possibilities may be placed on a continuum. One extreme is the model used in many universities in the British system, in which no courses are mandated in a doctoral program. The "research student" is expected to enter the program with a research agenda that is already quite well developed, and all courses attended are voluntary, for the purpose of intellectual development and increased expertise as it relates to the research focus. No examinations in courses need be taken; papers written are voluntary and again will probably be specific to the chosen research topic.

The other extreme of possible arrangements in a doctoral program is one in which all students in the program move together in a cohort through mandated courses, without personal choice of courses. In one instance, this arrangement may be necessitated by a cohort of learners in a distance-education program in which a large number of students start together, live at a distance from the institution offering the program, and expect to finish the degree together after a certain number of years. The expense of offering such courses, which may include site visits of faculty to a venue accessible to the students, and even a webmaster to manage the customization of a website for its purposes, may prohibit a proliferation of such courses, and may render such courses unworkable for the institution.

Most institutions offering doctoral programs in mathematics education will choose arrangements that lie somewhere between these two extremes. In the United States it is customary to require some course work in a doctoral program, often with a few advanced courses that form a core, supplemented by courses chosen by student and advisor(s) to customize the degree to the needs of the particular student. There are several issues and possibilities. The following account is based on the discussions of participants in two seventy-five minute sessions during the National Conference on Doctoral Programs. The questions addressed were, on the one hand, whether there should be a basic core of courses for all students in a doctoral program, and on the other hand, if the answer is affirmative, what courses should be included in such a core.

---

[1] *Norma C. Presmeg was at Florida State University when the conference was held.*

Issues

The following issues were discussed by participants on the road to reaching some conclusions. Without deciding whether a basic core of courses for all doctoral students is desirable, possible content of such a core was explored. How much mathematics content should there be in a doctoral program in mathematics education? Some institutions require the equivalent of a master's degree in mathematics for entrance into a doctoral program in mathematics education. Others specify a minimum number of semester hours of graduate mathematics course work (e.g., 18 hours at Florida State University) to be taken in the doctoral program if a student has not already satisfied this requirement at the master's level. How much mathematics education course work is desirable? It was recognized that proliferation of core courses could overload a program. The possibility of using a model of required processes and structures instead of core content was explored. Perhaps instead of a list of courses, agreement on a structure of core categories would be beneficial. For instance, one suggested structure included the following categories:

- mathematics learning;
- teaching and teacher education;
- mathematics curriculum.

Mathematics education research, it was suggested, could be woven into all three, or, alternatively, could be another category in this developing field. One model (Ohio State University) suggests "on the job" education experiences as students prepare to become mathematics educators: this could be called internship for the profession.

What exactly is the job for which doctoral students in mathematics education are being prepared? Is it that we are preparing leaders in the field of mathematics education and mathematics education research? One participant expressed the view that she did not feel like a leader five years ago when she completed her mathematics education doctoral degree: it was the start of an ongoing learning process. With different students considering several potential job markets, including "teaching" and "research" institutions as well as community colleges, the necessity for individualized programs of study was acknowledged. For those students who needed certification to enable them to teach mathematics in schools, what provision could be made in doctoral programs? The broad range of individual needs reinforced the view that a structure of core categories was preferable to a list of required doctoral courses.

Continuing in the vein of a core structure, it was asked whether there could be certain expectations when hiring someone who has a doctorate in mathematics education. In this regard, what are the sorts of understandings, skills, and dispositions that someone who receives a doctoral degree in mathematics education could be expected to have? Is there a need for generalist courses in mathematics education? A list of skills was not favored, but certain aspects of the various forms of the profession of being a mathematics educator were considered important enough to be required in a core structure. Some of these are as follows:

- Orientation to the profession.
- Political awareness.
- Internationalization and awareness of global perspectives.
- Preparation for reviewing and editing of manuscripts.
- Accomplishment in literary critiquing: students could be given the experience of preparing papers for and against particular positions, in order to help them see several viewpoints and think through issues.

One intriguing possibility that was explored was considering alternative forms of the dissertation. Instead of the usual 200-page or so dissertation, students might write two or more focused research-based papers suitable for submission to a journal (See Stiff, *Discussions on Different Forms of Doctoral Dissertations*, this volume).

What kinds of collaboration are possible in the preparation of mathematics education doctoral students? Rather than a relationship of dichotomy or antagonism, schools of arts and science should be viewed as working with mathematics education in this endeavor. Mathematics departments, too, are potential allies. Networking, dissemination, and joint projects are all regarded as valuable.

Participants finally came to the conclusion that it is beneficial to students to have a core in a doctoral program. However, this core may not necessarily be a set of core courses. It was suggested that there should be a set of core competencies, or standards, or a structure of core categories. At this point, decisions about the nature of this core are made according to the needs of departments and programs within the institutions housing doctoral programs and their students. Whether there should be a national core set of requirements remains an open question.

Norma C. Presmeg
Illinois State University
Mathematics Department
Campus Box 4520
Normal, IL 61790-4520
npresmeg@msn.com

Sigrid Wagner
Ohio State University
257 Arps Hall
1945 N. High Street
Columbus, OH 43210
wagner.112@osu.edu

CBMS Issues in Mathematics Education
Volume 9, 2001

THE TEACHING PREPARATION OF MATHEMATICS EDUCATORS IN DOCTORAL PROGRAMS IN MATHEMATICS EDUCATION

Diana V. Lambdin, Indiana University
James W. Wilson, University of Georgia

At the National Conference on Doctoral Programs in Mathematics Education, several sessions were devoted to discussion and debate about the teaching background and preparation expected of mathematics educators. The varied perspectives listed below were offered as points of departure for these discussions:

- Mathematics education doctoral programs should prepare *all* their graduates as mathematics teacher educators, a job that involves teaching both prospective and in-service teachers.

- Mathematics education doctoral programs may prepare *some* students for teaching mathematics at the college level, though not necessarily for working with teachers or prospective teachers.

- Mathematics education doctoral programs may prepare *some* students for positions with no teaching responsibilities, such as working in a research institute.

There was a consensus that teaching experience and preparation for future teaching are extremely important for mathematics educators, but beyond that, there was little agreement. Although teaching is one of the most intensive, time-consuming activities of a university professor's work, and although the public probably views teaching as the work that university faculty do, it is not clear that doctoral programs of study, in general, prepare junior faculty well for teaching responsibilities. Certainly, students in American institutions of higher learning do not necessarily bring any teaching experience with them to their doctoral studies. Indeed, in many university departments (e.g., mathematics, English, philosophy, chemistry) the typical student begins doctoral studies with little or no prior teaching experience (though some may have served as teaching assistants in some college-level courses or may have supervised laboratories or discussion sections during study for the master's degree).

Mathematics education doctoral students, on the other hand, are more likely—as are graduate students of education more generally—to have teaching experience prior to beginning their doctoral studies. This observation led participants at the conference to debate the importance of teaching experience prior to admission to a doctoral program in mathematics education.

TEACHING EXPERIENCE AS A REQUIREMENT FOR ADMISSION TO DOCTORAL STUDIES

In many cases, doctoral students in mathematics education have already had distinguished careers as school mathematics teachers before they enroll for doctoral studies. In addition to having taught children, they often have considerable experience conducting workshops for their peers and sharing teaching ideas at local, regional, or national meetings. Indeed, a 1998 survey of 48 of the top U.S. mathematics education doctoral programs (Reys, et al., this volume) indicated that K–12 teaching experience was a requirement for admission to 56% of the doctoral programs surveyed. Teaching experience was even more strongly emphasized in the smaller programs. For example, K–12 teaching experience was required for admission by 50% of the Group 1 programs (those awarding 25 or more doctorates from 1980 to 1997), as compared to 60% of the Group 2 programs (those awarding just 8–24 doctorates from 1980 to 1997) and 71% of the Group 3 programs (those awarding fewer than 8 doctorates from 1980 to 1997). Although there seems to be strong endorsement of K–12 teaching experience for mathematics education doctoral students, assessment of the quality of that experience has not been addressed. Indeed it seems to be assumed that any experience is good experience.

On the flip side of these data, we can observe that a sizeable number of mathematics education doctoral programs are apparently willing to admit students with no K–12 teaching experience (44% of all programs, and 50%, 40% and 29% of the Group 1, Group 2 and Group 3 programs, respectively). Some doctoral programs that admit students without prerequisite K–12 teaching experience may incorporate K–12 teaching experience or internships into an individual's programs of study. Such experience and internships can build upon the advanced study the doctoral student has encountered. Examples were cited of highly successful mathematics educators whose K–12 experience was acquired in alternative ways concurrently with doctoral study.

We know from the surveys and from job applications by graduates of some programs that doctoral students are admitted to study and allowed to complete their studies without K–12 experience. No one at the conference openly supported such a view. This might be rationalized for teaching positions in college mathematics departments, for work in research institutes, or for positions in regulatory organizations. Yet, applications showing no teaching experience frequently come for positions involving mathematics teacher education.

Recently, one author received a personal communication from a respected leader in mathematics education making the argument that excellence in research and scholarship was sufficient. The point was "a good researcher will become a good teacher." No one at the conference seemed ready to publicly endorse this view. Yet, the fact remains that with all the rhetoric in universities and colleges about criteria for promotion and tenure, the criteria for research productivity predominate. Excellence in teaching must continue to gain in importance as a criterion for promotion and tenure in order for performance in collegiate teaching to become more central to doctoral programs.

Many mathematics education programs have teacher education as their primary focus, and among those that do, it would be reasonable to expect K–12 teaching experience as a prerequisite for admission. On the other hand, mathematics education programs that focus more directly on research on teaching and learning, or on college-level mathematics teaching, might well have different expectations of incoming students or of their graduates. Some mathematics education doctoral students have no K–12 teaching experience, yet they bring significant junior college or university-level teaching experience with them to graduate school. (Unfortunately, the survey (Reys, et al., this volume) provides no data about programs requiring these sorts of experiences.) The largest mathematics education doctoral program, Teachers College–Columbia, seems to be a case in point. Most of their doctoral students are part-time students holding down post-secondary mathematics teaching positions in the New York area. If we consider doctoral students with post-secondary teaching experience together with those with K–12 teaching experience, it seems not only possible, but quite likely that the vast majority of mathematics education students come to their doctoral studies with some background in teaching at one level or another. Some conference participants were adamant that their program would not admit prospective doctoral students who did not have at least some previous teaching experience.

DIFFERENT TEACHING EXPERIENCES FOR STUDENTS WITH DIFFERENT BACKGROUNDS AND GOALS

The observation that doctoral students' teaching backgrounds vary widely raises questions about how any doctoral program can be expected to provide appropriate and varied teaching experiences for all students during their doctoral studies. The answer is that a program's expectations and offerings may differ (indeed, probably should differ) for students with varied backgrounds. For example, programs must provide different sorts of teaching experiences for:

- experienced elementary teachers,
- experienced middle school or secondary teachers,
- experienced junior college or community college teachers;
- transfers from graduate study in mathematics with only college teaching experience,
- students with strong mathematics backgrounds but no teaching experience,
- students from outside the United States.

Discussions during the National Conference on Doctoral Programs showed that opinions vary widely about what sorts of teaching experiences are appropriate to provide during doctoral studies. In fact, the discussions seemed to bog down over the recognition that students' career goals may differ as widely as their backgrounds. The majority of attendees were adamant that *all mathematics educators must be prepared as teacher educators*—not only knowledgeable about teaching and learning issues, but also competent in communicating with school teachers and administrators and informed about how schools function. Those who argued this point of view contended that the broader educational community expects an individual with a degree in mathematics education to be prepared as a teacher educator, regardless of whether the student plans to work as a teacher educator in the future or not.

One discussion at the conference led to a proposal that doctoral students should be expected to "move up or down one level from their past teaching experience." In other words, individuals who had taught grades 9–12 would be expected to get experience with the middle or elementary school level, as well as with beginning collegiate courses.

If a doctoral student was an experienced elementary teacher, then he or she would be expected to become conversant at least with middle school mathematics teaching, and perhaps with early childhood/preschool as well. Those who had taught only at the college level would be required to gain experience at least with secondary school teaching. It was not clear how strongly this proposal was supported, yet no one expressed disagreement with it.

A few people argued that a mathematics educator might not need background experience with teacher education at all if he or she intended to specialize in, for example, large-scale testing and assessment or in program evaluation. Others maintained that testing and evaluation experts need perspectives on teaching just as much as those who actually will work directly with students or teachers. Interestingly, Reys' survey (Reys, et al., this volume) showed that *preparation for teaching probably is very important for most mathematics educators*, since the majority of graduates from mathematics education doctoral programs take first jobs where teaching is a primary expectation (working in U.S. universities both small and large, in community or junior colleges, in K–12 schools, or in international universities). Very few graduates (fewer than 10%) take non-teaching jobs, and those primarily involve working for international, commercial or governmental entities.

There did seem, however, to be somewhat general agreement during the conference discussions that no doctoral program can be expected to be all things to all people and that program expectations and experiences typically must be individualized for students. This perspective fits with several findings from the Reys survey. When respondents were asked to describe features that distinguished their doctoral program from others, one of the notable claims was individualization (a program designed around the varied career goals and needs of individual students). This makes sense because doctoral programs in mathematics education typically do not produce large numbers of graduates each year, and due to their small enrollments, they cannot afford to offer many courses specially designed for their students. As a result, many of the professional experiences offered to mathematics education doctoral students are, by necessity, organized as independent studies, internships, or mentoring situations. This situation may be at least as true for teaching experiences as for any other sort of experience. Clearly, individualized teaching experiences differ widely, depending upon a student's career goals:

- teaching in a major research university vs. in a small college,
- teaching in a college or university mathematics department vs. in a school of education,
- specializing in elementary, secondary, or college mathematics education;
- serving as a curriculum supervisor for a school district,
- working as a policy analyst or a testing or evaluation expert,
- returning to one's country of origin to work in teacher education or collegiate mathematics teaching.

INDIVIDUALIZING TEACHING EXPERIENCES DURING DOCTORAL STUDIES

Discussion at the conference very quickly turned to questions about how it is possible to provide appropriate individualized teaching experiences for students with so many different needs and desires. One of the thorniest issues seems to be what to do when a student is admitted without significant teaching experience, yet needs teaching experience to reach his/her individualized goals. Proposed remedies for this situation included:

- having the student take leave from formal studies to teach in a school for a time;
- arranging for the student to enroll in a variety of teaching internships as part of his/her program of study. For school-based experience, such internship options might include student teaching, substitute teaching, or collaborating with K–12 teachers in multiple internship environments such as an elementary school, middle school, secondary school, urban environment, technical school, etc.;
- hiring the student for work as a teaching assistant or teaching intern under faculty supervision (for college-level experience). It is common practice for financial support of full-time doctoral students to come primarily from institution funds such as those for teaching assistantships, and this pattern holds for mathematics education doctoral students (Reys, et al., this volume).

Once a student has sufficient teaching experience to teach on his/her own, there are various college-level positions that provide invaluable preparation for the sorts of teaching the mathematics educator may be expected to do after graduation. These include:

- teaching one's own section of a course using a common syllabus (developed by a faculty member or a team of faculty members and graduate students);
- teaching a course based on a syllabus developed on one's own;
- teaching mathematics content or methods courses for elementary teachers;
- teaching mathematics content or methods courses for secondary teachers;
- teaching mathematics for nonteachers (e.g., developmental mathematics, pre-calculus, calculus, discrete mathematics, history of mathematics, college geometry, mathematics appreciation, etc.);
- supervising field experience students or student teachers in schools;
- conducting field experience seminars;
- teaching preservice teachers in integrated courses or courses not directly related to mathematics education (e.g., math/science methods courses, or courses in technology in education, diversity in education, or general methods);
- teaching research methodology courses and/or supervising undergraduate or graduate-level research projects.

Unfortunately, it is not always possible for a doctoral program to offer all of these types of experiences to students who need them. For example, a mathematics education doctoral program housed in a school of education might be unable to offer students appropriate experiences in teaching college-level mathematics if graduate students in the mathematics department in the college of arts and science are given priority in hiring, or if there are no faculty members in the mathematics department willing to serve as teaching mentors.

Because doctoral programs tend to be so intense, very few graduate students have time to obtain broad experience with the full range of teaching responsibilities that will probably consume much of their time and energy when they are later new faculty. As a result, many new faculty report that they are uneasy with the number of new preparations and variety of courses that they are required to teach during their first few years. In addition, recent curricular changes in undergraduate teacher education often include emphasis on multicultural, international, interdisciplinary, technology-based, writing-across-the curriculum, or service-learning aspects of education. New faculty may find that they are expected to develop courses incorporating these non-mathematics-specific issues even though they may have had no experience with them during their graduate teaching experiences. There are, however, a variety of not-yet-mentioned experiences through which graduate students can be better prepared for their more general teaching roles as faculty members. These include:

- participating in teaching assistant orientation and development programs;
- enrolling in general seminars on college teaching or on teacher education;
- developing a teaching portfolio (perhaps as part of a qualifying exam);
- working with a "teaching mentor."

Universities that have begun, in recent years, to pay more attention to teaching are likely to have begun to organize experiences such as these for their faculty and graduate students. In fact, when respondents were asked for their views on anticipated changes within their mathematics education doctoral programs in the next five years, there were a number who mentioned a new or increased emphasis on preparation for college teaching (Reys, et al., this volume).

RECOMMENDATIONS FOR DOCTORAL PROGRAMS IN MATHEMATICS EDUCATION

Doctoral programs in mathematics education must ensure that students are involved in a variety of teaching experiences, both in the schools and at the university level. These experiences should be carefully chosen to match the students' backgrounds and career goals. Experience with teaching should begin even before a student enters graduate school and should continue throughout the graduate studies. Students need more than just the experience of teaching new levels and types of classes, whether in collaboration with others or independently. They also require constructive feedback about their performance, and they can benefit from group discussions about creative teaching possibilities and problem solving.

The model used for training graduate students in research could be followed in building graduate students' competence and confidence in teaching and working with students. Departments or graduate schools may find it worthwhile to offer (general) seminars on teaching as a first step. Later experiences might include supervised teaching, team teaching, summer school teaching, and independent teaching fellowships after the more typical experience of teaching as an intern or assistant or leading discussion sections or supervising school field experiences. Finally, an interesting new development is Preparing Future Faculty (PFF)—a national program (the web address for which is www.preparing-faculty.org) organized to help colleges and universities develop programs for mentoring both graduate student teachers and new faculty.

It is very important for mathematics education faculty to help doctoral students learn how to document, reflect on, and evaluate their own teaching. This may be accomplished by such methods as involving students in examining their own teaching practices through a professional development model or requiring them to assemble teaching portfolios. Much is made of the "apprenticeship" concept for developing research expertise in doctoral programs in mathematics education. A parallel argument for an apprenticeship in collegiate teaching can be made. Assessment of successful completion of the research apprenticeship typically involves some combination of comprehensive examinations and compilation of evidence of research productivity, such as a portfolio of best work. Assessment of the successful completion of a collegiate teaching apprenticeship should be similarly considered. Use of a teaching portfolio for this purpose was discussed at the conference.

In sum, participants at the conference agreed that teaching must be an integral part of any mathematics education doctoral program, but there was lively debate about the range and diversity of the teaching experience expected of graduates of such programs.

Diana V. Lambdin
Indiana University
School of Education
201 N. Rose Ave.
Bloomington, IN 47405-1006
lambdin@indiana.edu

James W. Wilson
105 Aderhold Hall
University of Georgia
Athens, GA 30602-7124
jwilson@moe.coe.uga.edu

Discussions on Different Forms of Doctoral Dissertations

Lee V. Stiff, North Carolina State University

Consensus

Differing perspectives fueled the discussions about dissertations in mathematics education, but participants agreed that doctoral students should be able to synthesize and communicate research findings in the field. Great importance was given to doctoral students' ability to communicate with classroom teachers about mathematics education research and its usefulness in classrooms. Furthermore, participants felt that the need of doctoral students to conduct research beyond the acquisition of the doctoral degree was great. This was expressed in observations that pointed out that no matter what academic setting new doctorates in mathematics education may find themselves, expectations of establishing a research agenda are high.

Possibilities

What are possible ways of writing a dissertation or demonstrating research competency? The traditional five-chapter structure for dissertation dominated the discussions. It was generally agreed that while the 'completeness' of traditional dissertations promoted the conceptualization of ideas and relationships, as well as the concomitant collection and analysis of data, most participants observed that such dissertations did not necessarily facilitate doctoral students' ability to synthesize and condense research data and findings. Consequently, a critical aspect of research, the synthesis of information and integration of the results into an overall framework that advances the knowledge base, is often neglected in traditional dissertations.

Alternate 'dissertation' forms were suggested. Among them were (a) portfolios of several different research studies, (b) historical reviews, (c) innovative curriculum development projects, and (d) article dissertations. Each of the alternatives were discussed; however, much attention was given to the article dissertation format.

Article Dissertations

A journal-ready version of a traditional experimental dissertation was called an 'article dissertation.' Real interest was created during the discussion as several participants described the type of article dissertation in use at their universities. Although differences among article dissertations were evident, most article dissertations consisted of an introduction and statement of the problem, a literature review, and the journal-ready article, usually a 40-page APA-styled manuscript. Clearly, the perceived benefit of article dissertations was in the experience gained from synthesizing and focusing the research findings. Furthermore, the potential for widespread dissemination of the research is greatly enhanced in an article dissertation.

Differences among implementations of the use of article dissertations were many. Differences were easily conveyed by the following series of questions, the answers to which varied widely among participants. Should the article dissertation be submitted for internal or external review? To what type of journal should the article dissertation be submitted? Should both a research article and a research in the classroom article be written? Must the article dissertation be accepted for publication to be considered complete? Is there a need for data-filled appendices accompanying article dissertations? Are students at greater risk or disadvantage when writing an article dissertation?

OPPORTUNITIES

The conversations around the table clearly demonstrated that there is much interest in alternate dissertation formats. Participants were concerned that the dissertation plays an important role in preparing students to conduct research while orienting them to the expression of scholarship valued by the community of researchers in mathematics education. Everyone felt that the community of mathematics educators should collectively identify acceptable discourse for doctoral dissertations. Participants seemed to be in search of innovative ideas for demonstrating research competency in the written form. The need remains to discuss what is being done and what is yet possible.

Lee V. Stiff
North Carolina State University
326 Poe Hall, Box 7801
Raleigh, NC 27695-7801
lee_stiff@ncsu.edu

CBMS Issues in Mathematics Education
Volume 9, 2001

BEYOND COURSE EXPERIENCES: THE ROLE OF NON-COURSE EXPERIENCES IN MATHEMATICS EDUCATION DOCTORAL PROGRAMS

Glen Blume, The Pennsylvania State University

Doctoral programs in mathematics education typically consist of a combination of course work and a variety of outside experiences. Those non-course experiences (NCEs) often include activities that engage doctoral students in teaching, research and scholarship, outreach and teacher professional development, or some combination thereof. The activities may be formal or informal. Formal activities include teaching courses as a teaching assistant and learning about research while working as a graduate assistant for a research project. Informal experiences include editing or critiquing a document, preparing an article for submission to a professional journal, or observing exemplary teaching.

ASSUMPTIONS ABOUT NCES

In discussions of the nature and role of NCEs in mathematics education doctoral programs, faculty members indicated that they value such experiences and believe that their students value them as well. Several widely held assumptions about NCEs in mathematics education doctoral programs appear to provide the basis for such opinions. The first is that NCEs make *substantial contributions* to students' programs of doctoral study. Because of the potential for such contributions, a second assumption is that noncourse experiences should be an integral part of *every* student's program. A third, related assumption is that such experiences should be a *systematic* part of all mathematics education doctoral programs.

THE NATURE OF NCES AND THEIR ROLE IN STUDENTS' PROGRAMS OF STUDY

If NCEs are to be valuable experiences, their goals, nature, and means of implementation must be clear to faculty and students. The following sections address potential goals of NCEs, offer a list of NCEs perceived by the participants to be most important, and discuss issues relating to inclusion of NCEs in students' doctoral programs.

GOALS OF NCES

First, NCEs should complement course work. To do so, they should develop competencies that may not be addressed easily in the context of course work due to constraints imposed by time, the structure of course credits, assessment practices, and the like. An example of such an NCE is the design, conduct, and reporting of research as a result of long-term engagement with a research project, an experience seldom possible in the context of a semester-long course. Other examples are curriculum development experiences that require specification and initial writing, classroom-based piloting and associated research, and subsequent revision and field-testing of materials.

NCEs should develop students' expertise through experiences that approximate those encountered by mathematics educators who hold doctorates. Engagement in practical and authentic teaching, research, outreach, editorial, and networking experiences often takes place more readily outside the confines of courses. The flexibility and individualization afforded by NCEs often permit students to develop specialized competencies that might be unattainable through courses that must simultaneously meet the needs of many individuals.

Capabilities that might best be developed include expertise as a preservice and in-service teacher educator, ability to conceptualize and conduct research, fluent written and oral defense of one's work, ability to critique research and curricula, active participation in a scholarly community, familiarity with related disciplines (e.g., pure or applied mathematics or psychology), and expertise with technology. A wide variety of NCEs might contribute to the development of those capabilities.

"CORE" NCES

Based on NCEs that are currently available at, or deemed desirable by, various institutions, conference participants identified a collection of core experiences that might be sought in NCEs as part of a set of minimum expectations (Part 2) for doctoral students. Although participants did not intend the list to be exhaustive, they identified the following NCEs as important and desirable experiences for mathematics education doctoral students:

- Mentored teaching of mathematics (e.g., mathematics for elementary teachers);
- Mentored teaching of courses on the teaching and learning of elementary or secondary mathematics (often referred to as methods courses);
- Design of a course or development of curriculum materials;
- Development of technological tools for learning and teaching mathematics or development of expertise with such tools (or substantial capability to acquire such expertise);
- Mentored supervision of preservice teachers' field experiences;
- Development of a broad-based reading list that comprehensively addresses the field of mathematics education;
- Mentored conceptualization, conduct, and reporting of research;
- Development of writing expertise as might be evidenced by submission of an article appropriate for a practitioner or research journal;
- Oral presentation and defense of one's scholarly work (e.g., colloquium presentations or conference presentations that are practiced and subjected to peer critique);
- Service as a referee or an editorial assistant for a professional publication;
- Interaction with a local scholarly community as either a colleague or a critic (e.g., participation in collaborative projects, structured scholarly interactions in colloquia, or critiques of presentations);

- Interaction with the broader mathematics education scholarly community (e.g, through student-initiated contacts with other researchers or interactions at conferences, or contacts with visiting faculty); and

- Design, conduct, and assessment of long-term inservice or professional development activities for teachers (e.g, assistance with summer workshops and institutes or ongoing teacher professional development projects).

The preceding NCEs provide a variety of teaching, research and scholarship, and outreach opportunities that could prepare students for activities in which they likely would be engaged after completing a doctoral degree.

Other substantial NCEs occur in the context of assessments that are part of doctoral programs, such as the comprehensive examination and the oral defense of one's dissertation research. Participants agreed that the comprehensive examination assessed a student's ability to synthesize the knowledge and understanding obtained in his or her program of study, the ability to recognize and argue two sides of an issue, and the oral defense of his or her ideas. These goals also are pertinent to the oral dissertation defense.

SELECTION OF NCES AND THEIR IMPLEMENTATION IN STUDENTS' PROGRAMS OF STUDY

When discussing how NCEs might best be incorporated systematically into students' programs of study and how they might be made available to all students, conference participants agreed that expectations concerning NCEs must be clearly specified and conveyed to all students. These expectations might be expressed when the committee and student agree to include appropriate NCEs in a formal plan of doctoral study. They might then be restated in annual meetings of student and committee to assess attainment of the NCE goals in the plan that will culminate in the development of the student's portfolio. Some NCEs might be completed under direct supervision of a faculty member, while others might require little or no supervision. In some instances an appropriate array of NCE's might be used to define a program's residency requirement.

Some participants voiced concern that requirements for NCEs might unnecessarily "program" students into NCEs, thus constraining development of their expertise by prohibiting them from making choices that would reflect their evolving program of study and help to develop the professional competencies they perceived as necessary to their career goals. One proposed solution was that each doctoral committee prepare, in consultation with each student, a career path by NCE matrix, subject to revision in annual meetings as the student's career path evolved. It was further suggested that, since there exist a variety of career paths and since some NCEs would be more appropriate for some career paths than others, development of a prototype career path by NCE matrix would be informative to the field of mathematics education. Having clearly identified career paths, each of which would have a characteristic profile of recommended NCEs (as well as recommended courses), would be helpful across institutions for illuminating career options to students and tracking their participation in NCEs pertinent to their career path.

Another concern about NCEs centered on issues related to the absence of credit for NCEs. Some participants believed that students often see credit as legitimizing the effort they expend, and NCEs, by nature, generate no course credit. Although pressure to "get students through" their programs exists, that pressure should not be allowed to prevent students from including appropriate NCEs in their programs. Also, it is not only students who are concerned about the awarding of credit. Faculty members who supervise NCEs seldom get credit for such mentoring activities. The reality for many faculty members is that they mentor and supervise activities that extend far beyond what is considered an appropriate "load."

Inclusion of NCEs in students' doctoral programs appears to be widespread; however, their effectiveness is still in question. Do NCEs contribute to the development of students' expertise as mathematics educators, and, if so, how do they contribute? These concerns suggest the need for research targeted on NCEs.

RESEARCH DIRECTIONS

Despite rather universal acceptance of assumptions about the extent of the contribution of NCEs to students' programs, the desirability of all students engaging in NCEs, and the need to systematically include NCEs in students' programs, such assumptions are based primarily on anecdotal evidence. The field of mathematics education clearly must pose research questions related to those assumptions and use systematically gathered data to address those questions.

To determine the nature of the contributions of NCEs to students' programs, research studies should address questions in two general areas. The first area includes questions related to the nature and extent of mathematics education doctoral students' engagement with noncourse experiences. For example, how much of a typical doctoral student's program consists of what might be categorized as NCEs? Across doctorate-granting institutions what types of NCEs are most common? Which NCEs are required, and which are optional in students' programs? What is the relationship between the nature of students' career goals and the NCEs in which they ought to engage?

A second area includes questions that address the nature and extent of the benefits that doctoral students may realize from such experiences. What expertise results from NCEs? What are the various ways in which students benefit from NCEs? How does one measure the success of NCEs in achieving particular goals, and how successful are NCEs? Which NCEs are most useful to students pursuing particular career paths? Answers to such questions can inform the design of doctoral programs and help to maximize the impact of NCEs on the development of doctoral students' expertise.

Knowledge gained about the contributions of NCEs to students' expertise also might be used to improve credit courses in doctoral programs by incorporating some of the features of NCEs into more formal course work. More flexible and individual assignments and assessments and more innovative projects might result if successful practices from NCEs could be adapted and made available to more students within the context of credit courses. Flexibility, practical experience, long-term involvement, multiple iterations leading to the development of a product, and extended feedback all characterize many NCEs. Although it is likely that such characteristics contribute to the success of students' NCEs, confirmation thereof is necessary.

Another area of research concerns the sharing of information about NCEs. Institutions must systematically share information about their doctoral students' opportunities for NCEs and the success or failure of particular NCEs. Surveys of institutions and their graduates could generate data concerning the nature of successful NCEs and make such information more widely available to those planning doctoral programs of study in mathematics education..

A variety of research studies related to NCEs could provide much-needed information with the potential to improve mathematics education doctoral programs. If such information were available, doctorate-granting institutions might be more inclined to change their programs in ways that would provide optimal experiences for their students within and beyond formal course work.

Glen Blume
The Pennsylvania State University
269 Chambers
University Park, PA 16802
bti@psu.edu

PART 3: RELATED ISSUES

Organizing a New Doctoral Program in Mathematics Education

Carol Thornton, Illinois State University
Robert P. Hunting, East Carolina University
J. Michael Shaughnessy, Portland State University
Judith T. Sowder, San Diego State University
Kenneth C. Wolff, Montclair State University

Imminent retirement plans of a large proportion of the mathematics education faculty in this country, coupled with current and anticipated needs for strong mathematics education leadership, have created a great interest in and need for good doctoral programs in mathematics education (Reys, 2000). At the National Conference on Doctoral Programs in Mathematics Education, our panel was charged with considering issues related to starting a new doctoral program in mathematics education. Our recent experience in grappling with doctoral program initiation focused on six major aspects of this process and they will form the basis for our discussion.

Rationale / Key Parameters for Initiating a Doctoral Program in Mathematics Education

Although university settings and policies differ, a strong, coherent rationale is the assumed starting point for a proposal that seeks approval for a new doctoral program in mathematics education. This rationale might present arguments related to faculty strength, evidence of well-documented national and regional need—including the likelihood of attracting good students (both from the institution's own master's program and nationally)—cutting-edge features, or distinctiveness of the proposed doctoral program compared with other programs.

Longitudinal preparation is typically necessary. A university seeking to initiate a new doctoral program in mathematics education, for example, might build up faculty strength by focusing in its search for new hires on individuals who have graduated from strong mathematics education programs and have demonstrated their research capacity or their potential for research leadership. A core of capable faculty with active research interests who are willing to work with doctoral students is necessary to support the work of students in any doctoral program.

The National Conference on Doctoral Programs in Mathematics Education and, in particular, the status report on doctoral programs in mathematics education target the national need for mathematics education faculty in the United States today and in the foreseeable future (Reys, et al., this volume). The impact and reality of this need contribute to any rationale statement framed within the next few years for starting a new doctoral program in mathematics education. The fact that there is an acute shortage of doctorates in mathematics education is, in itself, a drawing card for students likely to consider new doctoral programs.

A strong master's program, either in mathematics or in mathematics education, is a distinct advantage, as it ensures the likelihood of doctoral students in a new program. Some educators caution against an individual taking both undergraduate and graduate work from the same institution. Nonetheless, a strong master's program often turns out candidates interested in studying mathematics education in greater depth, and such a program can strengthen the case for doctoral program approval.

A final element in a well-framed rationale for doctoral program initiation is the highlighting of cutting-edge features or elements that have the potential to make the institution distinctive as a doctoral-granting program in mathematics education. These features, idiosyncratic to each institution, may well include elements like emphasis on, and opportunity for, extended research in key content areas; cross-institutional, global, or cross-disciplinary opportunities; emphasis on teaching and teacher education, curriculum, or technology, including internet and interactive video teaching and learning possibilities; flexible scheduling; and a wide range of substantive out-of-course experiences.

In addition to careful framing of the rationale for a new program and characterizing of both its components and distinctive features, attention to institutional details like tuition waivers, graduate assistantships, requirements for class size, and departmental/ institutional policy regarding dissertation supervision is useful. These are among the aspects of program operation that contribute to eventual program success.

ADMISSION REQUIREMENTS

An examination of mathematics education position announcements makes it clear that institutions seeking to fill faculty positions in mathematics education want individuals with strong mathematics backgrounds. Individuals hired by mathematics departments to work at the secondary or early undergraduate level virtually always require master's degrees in mathematics, and universities seek at least a strong mathematics background for prospective faculty whose work will be with elementary teachers. Colleges of education similarly expect to hire individuals with strong mathematics backgrounds and, depending on the institution, sometimes require demonstrated master's level competence in mathematics. While not always formalized in a position announcement, a prevailing interest in K–12 teaching experience among potential hires has been revealed in surveys of institutions.

These facts have an impact on the admission requirements that an institution considers setting for entering doctoral students. Prior K–12 teaching experience appears to be a normal prerequisite for individuals intending to work professionally with K–12 teachers or in teacher education at these levels, though conceivably the doctoral program might be lengthened to include full-time classroom teaching. The master's in mathematics— particularly needed by those wanting to keep their options open for potential work as a K–12 mathematics educator in mathematics departments or top-quality departments of education—may be set as a prerequisite for entry or be part of listed deficits to be made up during doctoral study. The strong mathematics background for individuals seeking to limit their professional work in mathematics education to elementary and middle school students—sometimes interpreted as the equivalent of a bachelor's degree in mathematics with appropriate breadth—might likewise be listed either as a prerequisite for entry or as deficit course work to be taken while completing other aspects of the doctoral program. Other admission requirements, like letters of recommendation, written statements from applicants related to their professional goals or interests, GRE scores with minimum cutoffs, TOEFL scores for foreign-language students, or GPA during the last n hours of course work, are often specified by graduate schools or by individual departments.

PROGRAM FORMAT, COURSE WORK, AND OUT-OF-COURSE EXPERIENCES

Traditionally doctoral programs have been localized to a single institution, but potential advantages currently being recommended for cross-institutional efforts might suggest this consideration as well. In either case, doctoral students may be admitted either as part of a lock-step cohort group or on an annual or semester-by-semester basis. Some institutions allow a cohort group of students to complete (or nearly complete) the degree rather than admit cohorts on an annual basis. In practice, situations often arise that cause students in cohort groups to get "out of sync" with others—causing problems with course make-up. Many programs prefer rolling admissions of individuals on a semester-by-semester or annual basis over a cohort model. Institutions must really explore options and make these decisions internally.

General program requirements differ from institution to institution, but traditions in other universities provide guidelines for a beginning program. Most established programs involve a residency of at least one full year and advise full-time enrollment for as many other years as possible beyond this. As discussed earlier, a certain level of mathematics content work is required in addition to a core of research and mathematics education course work and ongoing seminars. Some programs require proficiency in a second language. Many programs require work or demonstrated competence in technology or supporting disciplines; most provide noncredit but significant out-of-course experiences. Such experiences, highly valued both by institutions and students, typically occur during full-time residency. Regarding these experiences, a careful balance between teaching (math content and methods for teachers); research, curriculum development, or grant work; student-teaching supervision in mathematics; and fieldwork with practicing teachers seems desirable. All programs culminate in comprehensive exams and dissertations.

Situations addressed by beginning programs differ because of the populations they serve, and some institutions make efforts to be as flexible as possible. Decisions related to many issues raised in this section are really local to the institution. They are influenced by the vision a department has for its doctoral program; by the overall quality and richness of experiences it wishes to provide; by staffing limitations; and by the size, make-up, and needs of the typical doctoral group.

MATHEMATICS EDUCATION COURSE WORK.

The substance of course work in core mathematics education courses at the doctoral level varies among institutions. Most doctoral programs include course work in mathematics education research, curriculum, mathematics education history, and issues related to K–12 teaching and teacher education. Many include course work that focuses on learning theories applied to mathematics instruction and issues related to teaching and learning-specific mathematics-content areas. Ongoing seminars allow students to learn about, reflect on, and analyze current, new, or emerging issues in mathematics education. When deciding on specific requirements for mathematics education courses, course topics such as these deserve special consideration.

ISSUES RELATED TO RESEARCH

It is generally accepted that doctoral students need substantive course work in both quantitative and qualitative research and that research experience beyond course work can be a useful and positive experience for students. This latter form of pre-dissertation research experience may be that associated with grants or professional projects built into the doctoral program. The depth and quality of the doctoral research experience appears to be directly related to the caliber of the course work, the richness of any pre-dissertation experiences, and the overall research strengths of the mathematics education faculty. Careful planning is needed when framing the research component and requirements of a new doctoral program.

DISSERTATION ISSUES

Several issues related to the writing of a dissertation deserve the attention of institutions developing new doctoral programs in mathematics education. Faculty research strength is critical for supporting dissertation work. Faculty time for directing a number of dissertations may be a consideration in program initiation and planning. To provide a forum for both faculty and peer interaction, most institutions offer a research seminar or course at the proposal-writing stage to support students as they frame their proposals for doctoral study.

A further, emerging issue is whether dissertations should continue in the traditional style or adopt the "article" format common in some European countries and several of the hard sciences in this country. This issue, discussed more generally by Krathwohl (1994) and more recently by Duke & Beck (1999), has also been recently discussed among mathematics educators. Instead of producing an archival document, the thought is to more carefully mentor a student toward actual publication of research in the culminating phases of doctoral study. It appears that current mathematics educators have real interest in the article dissertation format, and some are beginning to use it—a fact that emerging doctoral programs may want to consider.

Carol Thornton
Illinois State University
Mathematics Department
Campus Box 4520
Normal, IL 61790-4520
thornton@math.ilstu.edu

Robert P. Hunting
East Carolina University
Mathematics Department
129 Austin
Greenville, NC 27858-4353
huntingr@mail.ecu.edu

J. Michael Shaughnessy
Department of Mathematical Sciences
Portland State University
Portland, OR 97207
mike@mth.pdx.edu

Judith T. Sowder
Center for Research in Mathematics and Science Education
San Diego State University
6475 Alvarado Road Suite 206
San Diego, CA 92020
jsowder@sciences.sdsu.edu

Kenneth C. Wolff
Montclair State University
1 Normal Avenue
Upper Montclair, NJ 07043
wolffk@mail.montclair.edu

CBMS Issues in Mathematics Education
Volume 9, 2001

REORGANIZING AND REVAMPING DOCTORAL PROGRAMS—CHALLENGES AND RESULTS

Douglas B. Aichele, Oklahoma State University
Jo Boaler, Stanford University
Carolyn A. Maher, Rutgers University
David Rock, University of Mississippi
Mark Spikell, George Mason University

This panel presentation and the discussion that followed dealt with reorganizing and revamping doctoral degree programs at five universities. Since each of these programs is steeped in tradition, the process of change was particularly challenging. Although the programs are very different from most academic perspectives, panel contributors cited a common theme: declining enrollments. This problem is not unique to mathematics education; however, it is of concern to advanced graduate programs nation-wide. Prospective students are increasingly choosing to pursue other opportunities.

Another common theme is that existing programs reflect their departmental or institutional philosophy, which is based on a set of standards and expectations. Revamping an existing program in higher education is often perceived as "diluting" or "watering down" the program, and this is not received favorably. Change is difficult because it requires open-minded consideration of the context of the program. In this situation, for example, the amount of graduate-level mathematics required of graduate students is a major concern. How much mathematics should a doctoral student in mathematics education be required to complete? Is it a function of the level of the program (elementary, secondary, higher education)? (Dossey & Lappan, in this volume). Should there be a common core of mathematics for all doctoral students of mathematics education regardless of level (e.g., Should all graduate students be required to complete advanced calculus)?

One observes a great range in the amount of required mathematics in these five programs. Each presenter can offer a variety of reasons to justify the amount of mathematical content in his/her program. Setting aside the required amount of mathematics, the issue of the type of mathematics also warrants consideration; the mathematics curriculum nationwide has undergone significant changes largely due to advances in technology. Should what is regarded as classical mathematics be replaced by or augmented by technology-dependent mathematics? Each of the panelist's programs has wrestled with many issues. Common among the efforts to revamp the programs was the need to serve better the target audience (e.g., program level, location) and to redefine the program curriculum (e.g., mathematics, mathematics education, research). Also common among most of these reorganizations was either the replacement of the existing Ed.D. program with a Ph.D. program or the retention of the existing Ed.D. program with the establishment of a new Ph.D. program. These efforts to make the Ph.D. more available are in response to requests from potential doctoral students and employers to make the Ph.D. the degree of choice.

Although many possible organizational schemes could have been selected for this panel presentation, a format based on the level of the program (elementary, secondary, or postsecondary) was selected. An international perspective was offered by one panel member and emerged again in discussion after the presentations.

FIVE DOCTORAL PROGRAMS—HOW THEY ARE BEING RESTRUCTURED

A brief summary and discussion of each program as related to their changing structures follows. More specific information about these programs can be obtained directly from their institutions.

1. LEADERSHIP IN MATHEMATICS EDUCATION K–8, GEORGE MASON UNIVERSITY

George Mason University (GMU) was established in 1972 as an independent state-supported institution. Its development has been guided by the educational needs of the northern Virginia cosmopolitan constituency, located adjacent to Washington, D.C. In 1979 GMU began offering a doctor of arts in education (DA.Ed.) degree. This degree required joint acceptance by students in the department of education and a department (representing a specific content discipline) within the college of arts and science. The DA.Ed. was short-lived for several reasons. For one thing, the DA.Ed. was not well known to employers or potential students. Furthermore, education faculty wanted additional input in the planning and implementation of doctoral students earning a degree in education.

In 1993 GMU began offering a Ph.D. in Education. This Ph.D. degree program includes the study and practice of educational research as well as the study and application of theory and practice in a field of education. It also permits the equivalent of a minor in instructional technology. But, as a Ph.D. program in education, it retains a clear emphasis on and requirement for the type of research study and dissertation activity associated with a Ph.D. in education.

The program is individualized, interdisciplinary, and experiential. Students, with the guidance of faculty advisers, plan their own programs to meet self-defined goals. To accomplish these goals, students engage in a variety of intensive courses, independent studies, seminars, and internships of a highly practical nature. Success in the program requires a high degree of personal initiative, self-directed learning, and commitment to inquiry as a style of personal and professional growth. This Ph.D. in education degree has allowed interested faculty to develop and offer a mathematics education leadership program.

2. DOCTORAL DEGREE (PH.D.) IN SECONDARY (MATHEMATICS) EDUCATION, UNIVERSITY OF MISSISSIPPI

The secondary programs at the University of Mississippi were suspended in 1992 due to an NCATE review citing an inadequate number of faculty in secondary education. Recently the undergraduate and master's degree programs in secondary education have been reinstated; and a specialist degree has been added. These programs reside in the department of curriculum and instruction, which is housed in the school of education, and they have been carefully constructed with input from secondary education faculty as well as assistance from members of the faculty in the mathematics department. An important element in the restructuring of this doctoral program is its coordination with the qualifying standards for National Board Certification.

The greatest challenge has been resolving the debate of content versus pedagogy in the specialized-content area. The secondary education faculty (representing all subject areas) discussed the number of content courses that should be included in all the secondary doctoral programs. One obstacle to be overcome was the assumption that all of the secondary doctoral programs could be made to fit into a standard format. It became clear that the needs of doctoral students in each secondary education content area are very different. Therefore, a major programmatic challenge is to achieve a reasonable balance among content, pedagogy, curriculum and research courses.

The distribution of course work has not been resolved and continues to be discussed. It has been suggested that choices be allowed between several secondary mathematics courses, courses dealing with general secondary curriculum and instruction (such as those on special needs children and multicultural issues), related fields such as science education, and mathematics courses. This flexibility would allow a typical graduate of the program to provide quality work in several areas, depending on the student's professional goals and needs. The program will allow students room to explore avenues of research in mathematics education, to strengthen their content knowledge of mathematics, and to develop stronger approaches to preparing secondary mathematics teachers in the future.

3. DOCTORAL STUDY, RUTGERS UNIVERSITY

Rutgers University offers two doctoral tracks in mathematics education, an Ed.D. and a new Ph.D. The Ed.D. degree focuses on the building of a powerful understanding of the process of teaching and learning mathematics. It is anchored in close student reflection about how learners and others engaged in mathematical work build comprehension in detail and depth. Hence, emphasis on improving the learning and teaching of mathematics emerges from a deeper understanding of how people learn and use mathematics. Students conduct fundamental research in the psychology of learning mathematics and of solving mathematical problems. A bachelor of arts degree in mathematics and further study of mathematics at the graduate level are customary content background requirements for students.

Students in the doctoral program build a foundation to conduct their own independent research by participating, usually over three summers, in an intense practicum experience. Each practicum experience is centered on current research of faculty and/or other doctoral students. Student participation varies according to background and experience. The practicum work is complimented by a 12-credit research requirement that weaves through individual programs. Students have the option of also participating in practicum courses that are offered during the academic year.

Advanced students, those who have completed the practicum requirements and are working on particular research projects, serves as mentors to entry- and intermediate-level students, as appropriate. This cooperation creates an "inter-generational" experience for all involved. The practicum experience has become a basic and fundamental component of the doctoral program in mathematics education at Rutgers University.

Students come together to share in the planning, conducting, and analysis of ongoing research.

A new Ph.D. in education with a specialization in mathematics is being developed. The program requires a minimal study of mathematics to the master's level and focuses on the development of students' mathematical thinking, including their construction of models of mathematical concepts and processes, and their methods of mathematical problem solving; the impact of educational experiences on both the motivation to learn mathematics and the approach to that learning; and the social influences affecting mathematics education policies.

4. PH. D. IN MATHEMATICS (EDUCATION), OKLAHOMA STATE UNIVERSITY

Oklahoma State University (OSU) has a long-standing tradition of preparing college teachers of mathematics. Through cooperative agreement with the college of education, graduates of the Ed.D. program in the college teaching of mathematics have obtained positions primarily in departments of mathematics. Understandably, a benchmark of this program has been a strong content background in mathematics. Course work in education has been related to college teaching (e.g., critical issues in higher education, the community junior college, the academic department). Because of the nature of the target audience, this program is very different than those represented by other members of this panel.

In the program's early stages in the 1950s, college mathematics teachers completed their Ed.D. degrees with support from NSF fellowships. When NSF support for these fellowships waned, student recruitment relied largely on recommendations from former graduates. Motivated by the spirit of reform in mathematics education nationwide and the related changes in the demographics of college mathematics departments, and by the desire to have more control over the various components of the program, the mathematics department faculty has expanded its current Ph.D. program in mathematics to include a specialization in mathematics education. This difficult undertaking has required extending the customary thinking on the definition of the traditional Ph. D. in mathematics. The resulting Ph.D. program, which began in 2000–2001, as well as the existing Ed.D. program address the challenges facing mathematics departments in the future.

The Ph.D. in mathematics, with specialization in mathematics education, includes completion of the doctoral core mathematics courses, mathematics education courses (higher education and/or common education), research in mathematics education courses, examinations, and a major thesis. Built into these general areas, there are also unique components related to professional development, including the following: .

- A written professional development plan outlining the activities that the student intends to use to satisfy these requirements must be presented for evaluation as early as possible in the student's graduate career.

- A final professional development portfolio containing documentation and summary reports on each of the elements constituting the professional development portion of the degree requirements must include evidence of having met these requirements: execution of a curriculum development activity (e.g., write and field-test class activities for a course being taught; work in connection with a funded curricular project taking place within the department; development of projects for use with a course; write and field-test computer laboratory activities for a course being taught.); development and execution of two professional education presentations (e.g., state, regional, national meetings; teacher workshops; department seminars).

- Plan and carry out a two-year teaching assignment designed to provide a broad range of experiences with various teaching modes (e.g., reformed, traditional, calculator-based, project). Participate in the department's mathematics education seminar for two semesters and give at least two presentations in this seminar.

5. PH.D. IN MATHEMATICS EDUCATION, STANFORD UNIVERSITY

Doctoral programs in Stanford University's school of education focus on the development of high-quality research. Students are encouraged to conduct rigorous investigations of educational issues, drawing from a wide range of disciplines. Students of mathematics education work within the curriculum and teacher education (CTE) program. All CTE students take core studies addressing the role of teaching, curriculum, assessment, and evaluation. Mathematics education students are not required to take courses in mathematics, but they are expected to have a strong background in mathematics and at least 2 years of experience working in education (usually teaching). All doctoral students are required to choose minors in departments outside the school of education.

This presentation was unique in that it brought an international perspective to the restructuring of the program. The English presenter offered her background as a basis for considering the restructure of the Stanford University program. This perspective was particularly enlightening and added much to the discussions that followed. In England students do not attend any courses as part of the Ph.D., although some universities offer intensive workshops on research methods. The English Ph.D. process begins with a student posing his or her research questions, which he or she then pursues for a minimum of 3 years. The average time to completion in England is 7 years, which takes into account part-time and full-time students. Students are expected to learn about relevant research, theory, and practice in their field through extensive reading, which may be structured by the Ph.D. supervisor. The Ph.D. supervisor, while similar to the advisor in the United States, generally serves as the only academic contact for students, as the English process does not include the establishment of a committee. The quality of the Ph.D. experience in England is somewhat idiosyncratic—due to the importance of the student-supervisor relationship and the variability in training and support offered by universities.

Among the characteristics being considered in the restructuring of the Stanford doctoral program in mathematics education are

- awareness of the range of research methodologies and methods employed by researchers;
- training in research methods through the analysis of data;
- time to conduct in-depth and longitudinal studies;
- a close relationship with a mentor; and
- freedom to pursue the student's own field(s) of interest.

The presenter maintained that the Ph.D. is the highest academic qualification that may be conferred on any candidate, and it is, distinctively, a research degree. As such, the candidate should have performed high-quality research in his or her field and contributed to the generation of new knowledge. The dissertation should, therefore, be the defining aspect of a program. These two characteristics, which at times become lost in the concern for job preparation, are at the center of the restructuring work at Stanford.

Douglas B. Aichele
Department of Mathematics
Oklahoma State University
Stillwater, OK 74078
aichele@math.okstate.edu

Jo Boaler,
Stanford University
School of Education
Stanford, CA 94305-3096
joboaler@leland.stanford.edu

Carolyn A. Maher
Rutgers University
10 Seminary Place
New Brunswick, NJ 08901
cmaher@rci.rutgers.edu

David Rock
University Of Mississippi
School of Education room 152B
University MS 38677
rockd@oldmiss.edu

Mark Spikell
Graduate School of Education MS 4B3
George Mason University
Fairfax, VA 22030
mspikell@wpgate.gmu.edu

CBMS Issues in Mathematics Education
Volume 9, 2001

RECRUITING AND FUNDING DOCTORAL STUDENTS

Kenneth C. Wolff, Montclair State University

In an ideal setting, recruitment of candidates for doctoral programs in mathematics education would require only some minimal outreach. The reality is that successful recruitment is but one important component of a healthy doctoral program. This paper summarizes the results of a discussion group that met at the National Conference on Doctoral Programs in Mathematics Education. The discussion group expanded the original topics of recruiting and funding to include retention and apprenticing. This paper includes related information and comments from other conference sessions as well as some obtained outside of the actual conference. The paper *Doctoral Programs in Mathematics Education: A Status Report* (Reys, et al., this volume) provides important data about many aspects of doctoral programs in mathematics education. That paper together with additional information shared at the conference highlighted not only the need for mathematics education faculty, but also the wide range and targeted audiences of doctoral programs in the United States. Assuming that the national need for faculty in mathematics education will result in an increase in both the number of mathematics education programs and the number of interested candidates, the successful matching of candidates and programs becomes an important concern.

RECRUITMENT

Recruitment practices for doctoral candidates in mathematics education programs vary from institution to institution, are often a function of location as well as the intended audience, and include both traditional and newer, technology based approaches. Thus, in many ways, recruitment of graduate students in mathematics education shares characteristics associated with the recruitment of undergraduates into the field of education described in *Educational Renewal: Better Teachers, Better Schools* (Goodlad, 1994).

The traditional, time-honored methods of networking and word-of-mouth outreach still prevail, and the results of such recruitment and placement are usually satisfactory for both candidates and programs. Advertising in national journals and the broad mailing of program announcements to other institutions continue, but the efficiency and success of these outreach efforts is difficult to document. One easy-to-implement recommendation to advertising programs is to have the resulting inquiries directed to a return address that discloses the source of the inquiry. A few years of gathering data should help with the decision of how to allocate recruitment funding most efficiently. An important new medium for dispersing information about doctoral programs is the internet via the development of informational webpages. A webpage should, of course, be well designed, informational, and easy to read and navigate. This is especially true for the initial few pages, which should contain links to pages with more detail about items such as admission requirements, the application process, financial support, faculty and their research interests, and/or information about current students and recent graduates. If the webpage contains an e-mail contact, it is very important that the response be personalized to each inquiry and not just return an electronic form of the application and/or a repeat of the information on the webpages.

Programs associated with funded projects that support graduate students while giving them research, professional development, or similar opportunities reported that this aid was an important aspect of their successful recruitment efforts. At least one such program cautioned that it is important for potential candidates to clearly understand the difference between the goals of such projects and the goals and requirements of the program of study in mathematics education. Programs of study located in densely populated areas and designed for classroom teachers and supervisors should consider advertising in appropriate regional professional journals and newsletters and also in those of local bargaining agencies. Recognizing that alumni are an excellent recruitment tool, try to have your program mentioned in alumni publications and department newsletters. Maintaining an information booth at regional conferences and also at graduate school information fairs was also recommended. It was noted that some universities have graduate student recruiters, and at such institutions it is very important to work closely with those recruiters and keep them well informed about your program.

Although opinions vary as to whether it is in the best long-term interest of students, the practice of internal recruiting from a university's own master's and bachelor's degree programs should at least be considered. Another recommended recruitment technique was to develop a relationship with potential feeder schools by having selected faculty or doctoral candidates occasionally present at a seminar or student mathematics organization. This approach was particularly recommended for recruitment at undergraduate institutions with large minority-student populations. Informational letters describing a program should be directed to key personnel at locations with employees who might be interested in pursuing doctorates in mathematics education. Sending such letters to deans and department chairs or advisors at two-year colleges, and to supervisors and department chairs of regional school districts was recommended.

FUNDING

Doctoral study is expensive, and many students require some form of financial support. Such support is primarily available to those who pursue the degree on a full-time basis. The survey data reported the main source (53 percent) of support for full-time doctoral study is provided by institutional funding for graduate assistantships (Reys et al., this volume). While the maximum funding support was $32,000, the median was about $10,000 per year. It was noted that funding above the $10,000 level was often obtained by the packaging of grant funds, such as an assistantship supplemented with scholarships. Working for a local school district in a part-time position, often under the supervision of a university faculty member, was a recommended source of support. However, it was noted that such opportunities are sometimes limited by the language of local school contracts. In such situations it is sometimes possible for the school to hire the university as an outside contractor for such instructional services. In turn, the university then hires graduate students to deliver the services under faculty supervision. Other avenues of support included teaching or tutoring mathematics or statistics courses for other departments, and working as instructors in the mathematics department, positions which would include salary and benefits. Concern was expressed that the work performed in connection with the assistantship, grant, or other university-sponsored work should be enriching to the candidate's program of study. The work should also be reasonable in that it should not detract significantly from the time spent on graduate study. Fifteen hours of work per week was estimated to be a reasonable amount of time in return for a graduate assistantship. Some universities have strict guidelines or TA unions that govern such work assignments. Oklahoma State University has an innovative program in which the student who is a TA does not have any first-year teaching duties. Instead, he or she is assigned a mentor who helps with development of teaching skills and provides limited teaching opportunities in the mentor's own classes.

Other than assistantships, grants, and scholarships, funding to support graduate students varied. Funding for part-time students, when available, is usually in the form of scholarships, tuition waivers, and small stipends for the purchase of books. Georgia has a large number of scholarships funded by the state lottery; some universities have competitive block grants; and others are permitted to use salary savings to support graduate students. Asking alumni to tag their donations to support graduate students has been successful at some universities.

RETENTION

Although location is often the primary factor for a candidate's selection of a doctoral program of study, the quality of the program and how it meets the perceived future goals of the potential candidate is the primary retention-related variable that faculty can most readily influence. Funding support, such as scholarships tied to state lotteries as discussed in the preceding section, are not in the purview of program faculty. However, since funding support for graduate study is an important, often decisive, variable in program selection, it is important for faculty to be active in some aspects of funding as discussed in the previous section. How each faculty and program approach this activity is best decided at the department or program level.

Cohorts of doctoral students were recommended as a way of building a sense of community and commitment among candidates. Although this sense of community does develop rapidly, the cohort model is often difficult to maintain both because of the individuality of some programs of study and because of outside influences, such as illness or unforeseen family commitments. However, it is usually possible to provide a common set of course experiences for two or three semesters that will provide a cohort environment.

Small amounts of support and recognition can contribute to program satisfaction and retention and, in turn, to future recruitment. Such support and recognition might take the form of travel grants to attend an appropriate seminar or professional conference, small stipends to help defray the costs of books and research expenses, monthly seminars with light refreshments, an annual picnic, and a variety of small recognition awards to individuals for exemplary work in a particular course or project. Alumni, local businesses, or even university funds can sometimes be used to support such activities.

Good advisement by faculty members, which includes at least one face-to-face meeting per year with each student is important. Besides discussing programmatic concerns, the candidate's professional goals and progress toward them should be addressed. Faculty must be aware of student concerns, what is working, and what is not.

Exit interviews conducted by someone outside the department are a recommended form of feedback. In the case of programs with part-time students exit interviews could be revised to serve as a progress interview every two or three years.

APPRENTICING

Some conference attendees stressed the importance of maintaining contact with successful candidates as they begin their academic careers. By treating them as apprentices, especially as they develop individual programs of research and adjust to life after graduate school, faculty increase their potential for success. Perhaps there is a way to build upon successful efforts such as Project NExT (New Experiences in Teaching), which provides for mentors from other institutions who support beginning mathematics instructors, including mathematics education faculty.

Kenneth C. Wolff
Montclair State University
1 Normal Avenue
Upper Montclair, NJ 07043
wolffk@mail.montclair.edu

CBMS Issues in Mathematics Education
Volume 9, 2001

THE USE OF DISTANCE-LEARNING TECHNOLOGY IN MATHEMATICS EDUCATION DOCTORAL PROGRAMS [1]

Charles E. Lamb, Texas A&M University

Evolving technology has opened the gates for distance-learning opportunities. Although few participants had first-hand experiences with distance learning, there was keen interest in sharing and learning of ways to use this technology and in exploring opportunities for incorporating distance learning into doctoral programs in mathematics education.

A doctoral cohort program operating at Texas A & M was briefly described. The program involved 9 students from Brownsville, Texas who pursued their doctorate using distance technology. Several courses were delivered using interactive television, while others were taken in residence. Students and faculty communicated via phone, mail, and e-mail.

All students were able to complete their course work while working full- time jobs. Five of the students have progressed through the exam stage, and two have finished proposals and collected data for their dissertation. While the program has been successful, there were many obstacles to overcome. For example, teaching by TV is very different from teaching a class having direct personal contact with students. It challenges the professor to do things differently. There were also technical glitches that disrupted hookups, and a 3-second time delay between transmission of voice and picture that was disconcerting. In addition, there were times where e-mail was hindered by a downed server.

Norma Presmeg then shared experiences relative to a master's program between Florida State University and the Dade County Schools in Miami, Florida, which relied on distance learning. She reported the challenges of dealing with a large number of students and the increased faculty workload created by the volume of e-mail traffic that results from the personal communication with students that is a vital component of distance learning.

These two experiences led to a lively discussion of a range of issues. Among the observations made and questions raised were the following:

---

[1] This paper is based on discussions at a session on distance learning. No formal paper was prepared prior to this session, which provided a freewheeling discussion of the topic. Thanks to Charles E. Lamb for his effort in capturing the spirit of those discussions. This discussion of distance learning led to the solicitation of the paper by Lesh, Crider and Gummer that follows.

GENERAL OBSERVATIONS

- Many more potential doctoral students can be served via distance learning.
- Technology is changing every day. This is both a blessing and a curse, but it will provide many ways to connect to doctoral students in distant sites.

TEACHERS

- Professors need to accept the fact that teaching effectively via distance learning requires different teaching styles.
- Subtle issues such as dress, tone, vocal quality, etc. must be taken into account.
- Measures need to be taken to insure that all students feel they are a part of the class.
- Special training in the effective use of technology (i.e., ELMO) should be part of course preparation.
- Staff support for distance learning may require technical assistance to initiate electronic communication, as well as site monitors.
- Discussions about teaching load may need to be negotiated, particularly when courses are made available to students from different institutions.

STUDENTS

- What measures need to be taken to prepare students to effectively engage in distance learning courses?
- What characteristics are associated with students successful with distance learning?
- How are students evaluated? What do students think of this technology?

COURSES

- Do certain courses lend themselves to being delivered via distance learning?
- How should the organization of a distance course differ from an on-campus course?
- What copyright issues, as well as other legal issues such as ADA, are involved with distance learning?

Obviously, the discussion raised more questions than it provided answers. In conclusion, the vision here is very different from what we all went through in our doctoral studies. Will the doctoral students in a program with a heavy distance-learning component be better than on campus programs? Perhaps the safest answer is that these students will be different!

Charles E. Lamb
Department of Teaching, Learning and Culture
Texas A&M University
College Station, TX 77843-4232
celamb@tamu.edu

EMERGING POSSIBILITIES FOR COLLABORATING DOCTORAL PROGRAMS

Richard Lesh, Purdue University
Janel Anderson Crider, University of Minnesota
Edith Gummer, Oregon State University

With the emergence of Internet II and associated advanced technologies for communication and computation, new possibilities are becoming apparent that could revolutionize graduate education and research in mathematics education, science education, and technology education. This paper briefly describes initiatives now underway as part of a rapidly evolving Distributed Doctoral Program (DDP) that is jointly sponsored by Purdue University, Indiana University, Purdue-Calumet, and IU/PU-Indianapolis and that offers collaboratively taught courses involving leading researchers and faculty from programs throughout the United States and other countries. [1]

The DDP is a federation in which graduate students on any of the collaborating campuses have access to faculty advisors, critical courses, clinical experiences, and research and development projects on all campuses. The program is possible not only because of new distance- learning capabilities associated with Indiana's regional hub for Internet II, but also because of well-established cross-campus enrollment procedures established by the Big Ten's Committee on Institutional Cooperation (CIC) Traveling Scholar Program.

This paper has four main parts. The first describes problems and purposes that Internet-wise collaborating doctoral programs might be designed to address. The second describes some significant ways that relevant courses can be expected to change when they are taught using Internet-based collaborations among students and faculty members at multiple campuses. The third describes benefits that are perceived by participating faculty and students. The fourth describes some important organizational issues concerning administrative and technical support that should be considered when undertaking the design of a distance learning course.

---

[1] *Recent collaborators have included students and faculty members at SUNY-Buffalo, Rutgers, Syracuse, Arizona State, and the University of Minnesota—as well as Queensland University of Technology and the University of Quebec at Montreal in Canada.*

WHAT PROBLEMS AND PURPOSES CAN DISTRIBUTED DOCTORAL PROGRAMS BE DESIGNED TO ADDRESS?

In fields such as mathematics education, doctoral students represent classic cases of geographically distributed students with highly specialized needs, many of which are unlikely to be met unless some form of distance education is used. In spite of this fact, however, distance education has not inspired much enthusiasm among mathematics educators—mainly because the limitations of available media have forced it to rely on transmission-oriented forms of instruction that are antithetical to the views of teaching and learning that mathematics educators want to promote. For example, using carefully crafted television programs for children as models of excellence in distance education, videotaped lectures have been transmitted to students at remote sites, even though the live versions of these lectures often were of questionable quality and were not improved by their being filtered through technical media or experienced at remote sites. Digital libraries of such "lessons" have become common fare simply because, for students who otherwise might have no opportunities for learning in some topic areas, providing some opportunities seems preferable to providing none—even if the quality is questionable.

What's new today is that the availability and affordability of high-speed, interactive, dynamic, sharable, multimedia capabilities are enabling collaboration among diverse teams of students and mentors who can co-construct complex artifacts, ranging from multimedia presentations, to interactive simulations, to sharable and reusable conceptual tools. For example, in undergraduate teacher education courses, prospective teachers can use combinations of e-mail, voice-mail, and computer-based videoconferencing and file sharing to serve as: (a) tutors to diverse populations of students working on homework, (b) consultants to teams of students working on complex projects, (c) apprentices to teachers developing multimedia experiences for students in their classes, or (d) assistants to teachers analyzing and assessing students' work on projects such as those involving compositions, constructions, or proofs that otherwise would have been avoided because meaningful feedback would have been too complex to generate.

For Ph.D. programs in mathematics education, cohorts of Ph.D. students even at leading research universities are seldom sufficiently large to justify offering many highly desirable specialized courses (or internship experiences). Why are these specialized courses especially important today? Consider the following:

- During the late twentieth century, mathematics educators made enormous progress that shifted beyond theory *borrowing* toward theory *building*. Rapid increases occurred in the volume and sophistication of mathematics education research, and these increases ushered in a series of paradigm shifts that provided new ways of thinking about the nature of students' developing mathematical knowledge and abilities, as well as new ways of thinking about the nature of effective mathematics teaching, learning, and problem solving. These paradigm shifts also led to new research designs, assessment designs, and curriculum designs that are distinctive to mathematics education and that are not likely to be addressed in courses taught by and for generalists. Yet, even if a given university has enough mathematics education doctoral students to justify specialized courses about research design, assessment design, curriculum design, or software design in mathematics education, it is seldom the case that it has faculty members with expertise in all of these areas. Furthermore, similarly specialized courses that focus on distinctive characteristics of problem solving, learning, teaching, or teacher education in mathematics education may also be desirable.

• Beyond theoretical models and perspectives that should become familiar to mathematics education doctoral students, outstanding programs also should provide access to a wide variety of internships (or other field experiences) that should be available in diverse settings that range from urban to rural, from preschool through college, and from schools to museums, community centers, and business organizations. Furthermore, in these settings, students should have the opportunity to encounter mathematics educators who are leaders in curriculum development, teacher development, or program development, as well as in knowledge development (or research).

Both of the preceding problems involve finding ways to provide depth and diversity of experiences in highly specialized topic areas. Corresponding opportunities occur because, when new communication technologies are used in innovative ways, new resources often emerge from formerly untapped sources. For example, the PU/IU Distributed Doctoral Program was designed to deal with the following kinds of common problems and opportunities:

• Neither Purdue University nor Indiana University is located in a large urban area of the type that must be involved in many federally funded research programs. Consequently,, if either is to compete for funding, it must establish strong working relationships with satellites in areas such as Indianapolis, East Chicago, or Calumet. Also, for faculty members at campuses such as IU/PU (Indianapolis) or Purdue-Calumet, competitiveness for grants often is enhanced by strong working relationships with Research 1 Universities.

• There are mathematics educators on the faculty at both IU/PU (Indianapolis) and Purdue-Calumet whose credentials would qualify them to be senior faculty members at Research 1 Universities, and for campuses such as IU/PUI and Purdue-Calumet to attract and keep talented faculty members it is important that they be able to provide many of the kinds of research and doctoral advising opportunities that are available to faculty members at Research 1 Universities.

• There are mathematics educators working in school districts, on state boards of education, at regional museums, in regional research laboratories, and in nonprofit research and development institutions who are able to provide internships and serve as mentors to outstanding graduate students. Furthermore, if these organizations don't collaborate with universities, they often have difficulty attracting high-quality assistants to work on short-term, high-impact projects, and they often have difficulty keeping permanent staff members without providing continuing professional growth opportunities of the type that are available at leading research universities

• Any of the preceding institutions may have resources that others lack. These resources may range from special kinds of technical support (e.g., for configuring computers and computer networks so that videoconferencing works smoothly) to support for outreach activities in school districts with special needs. Furthermore, beyond benefits that come from sharing resources, collaborators often find that their "lobbying power" is stronger on their own campuses if they are part of a larger group.

Beyond creating new possibilities for specialized courses and field experiences, and beyond offering opportunities for both students and faculty members to build strong mentoring relationships with more diverse kinds of experts in a variety of fields, another reason why participants have been enthusiastic about Purdue and Indiana University's Distributed Doctoral Program is that it encourages community building—for both students and faculty members. For faculty members, this often focuses on the chance to embed their work in "something bigger" being carried on by a community of people with sufficiently shared perspectives to develop an identity that is capable of attracting resources (e.g., students, colleagues, technical capabilities, funding) to support the shared agenda. For students, it often focuses on building working relationships with a larger body of talented students with shared perspectives but different backgrounds of knowledge and experience.

A common criticism of mathematics education research is "It doesn't answer teachers' questions!" But, the view that "teachers should ask questions and researchers should answer them" is quite naïve, and collaborative research programs have the potential to deal with shortcomings associated with this point of view.

- Most of the challenges and opportunities that mathematics educators confront are sufficiently complex that they are not likely to be addressed effectively using results from a single research study. Rather than thinking in terms of a one-to-one match between research studies and solutions to problems, one must generally gather results from many research studies, which all must contribute to the development of a theory (or model) that evolves over a prolonged period of time—with input from many people and perspectives. That is, research contributes to theory, and theory contributes to practice.

- No clear line can be drawn between researchers and practitioners; there are many levels and types of both researchers and practitioners, and the process of knowledge development is far more cyclic and interactive than is suggested by one-way transmissions in which teachers ask questions and researchers answer them.

- Teachers aren't the only ones whose actions and beliefs influence what goes on in mathematics classrooms. For example, other influential individuals include policymakers, administrators, school board members, curriculum specialists, textbook authors, test developers, teacher educators, and others whose knowledge needs are no less important than those of teachers. Furthermore, for any of the preceding practitioners, the "problems" that they pose often focus on "symptoms" rather than underlying "diseases." So, what they ask for isn't necessarily what they need, and useful tools and conceptual systems usually must be developed iteratively and recursively—with input from many people over long periods of time.

- Productive knowledge development projects also often involve some form of curriculum or program development, and productive curriculum and program development projects often involve knowledge development. For example, during the past quarter century, if any progress has been made in projects aimed at curriculum development, software development, program development, or teacher development, it's precisely because more is known.

- In mathematics education, many people who are known as leading "researchers" also tend to have equally strong reputations as teachers, teacher educators, curriculum developers, or software developers. And the reverse is also true; many people who are best known in these latter areas are also highly capable researchers.

What all of these observations suggest is that research must be a great deal more cyclic, interactive, and cumulative than is often portrayed by those whose first and last pieces of research are their doctoral dissertations. Consequently, from the perspective of both students and faculty in the DDP, one of its most important goals is to encourage participation in an ongoing research community. In fact, we've even been able to encourage collaboration from science educators, technology educators, and others outside of mathematics education—at the same time that it is clear that not every mathematics educator on every campus needs to participate.

AN EXAMPLE OF A PILOT COURSE USED BY PURDUE UNIVERSITY SCHOOL OF EDUCATION

One of the first courses taught in the Distributed Doctoral Program engaged course participants in the exploration of research designs in mathematics and science education. It involved the regular participation of students and faculty at eight institutions, including two English-speaking international locations. There were 27 students enrolled in the course, and six faculty members also participated. Additionally, guest speakers and content experts from remote sites also participated in the course on occasion.

The participants met synchronously for one hour each week through a low-cost Internet-based videoconference. Discussions continued online throughout the week in both synchronous and asynchronous media including text chat, bulletin board discussions, e-mail, and a listserv. The course technology facilitator developed a syllabus for the course that used the WebCT courseware package as the platform within which students could access the various discussion forums, reading materials, and anonymous postings of class assignments for review and critique. The software used for the videoconference was White Pine's Meeting Point conference server that uses CU-SeeMePro, also from White Pine, as the client. Each site logged into the Meeting Point server over the Internet using CU-SeeMePro. The conference server was host to a maximum of 25 participants in a single conference session. All 25 participants received audio and video and were able to contribute through text chat, application sharing, and white board; up to 12 of those 25 could send video. The course regularly provided conferencing connections among fifteen to seventeen computers on any given week, with ten or twelve sending video.

One of the goals of the course director was to find a low-cost videoconferencing package that worked reasonably well over high-speed Internet connections rather than using costly dedicated lines usually associated with videoconferencing or interactive television. While the Meeting Point conference server runs in the $10,000 range, the CU-SeeMePro client that each site needs retails for $69. For this course, the host university supplied each site with one license for CU-SeeMePro. Sites with multiple computers logged in to the conference purchased additional licenses.

The text for the course was a new compilation of research designs in mathematics and science education coedited by the faculty member who determined course content and activities (Kelly & Lesh, 2000). Chapters from the book were posted to the WebCT site, and each week students and faculty at the various sites were assigned to moderate discussion of a particular section of the text. Several of the faculty members who participated in the course from different sites were chapter authors who provided additional insight into and expertise in the discussions. Students were required to contribute to the asynchronous discussion forums; engage in synchronous chat sessions with other graduate students and visiting "experts;" write introductory, three-page research proposals that were later expanded to larger, more detailed proposals; and engage in the weekly one-hour Internet videoconference.

A "buddy" system was established at the beginning of the course in which students from the host university provided initial welcoming and support for participants at distant sites. This personal element facilitated communication about the goals and procedures of the course. Later the system was modified to partner students with common research interests to facilitate substantive discussions about the content and issues of the course.

Certainly, not every course in mathematics education is suited for a technology-based distributed environment. This course was selected as the pilot course because it was a special topics seminar that might not have attracted sufficient enrollment to run as a conventional course. This is just the sort of course that is appropriate for a collaborative partnership such as the DDP. Also, the philosophy behind the program is to capitalize on the core competencies of faculty at various institutions—a pooling of the professorate. Accordingly, special topics courses germane to the talents and specializations of the participating faculty—courses that otherwise would be offered infrequently or not at all—are well suited to distributed programs. Core and other required courses may be less likely candidates for sharing—if these courses meet required minimum enrollment figures and are less likely to capitalize on the core competencies of faculty at other locations.

CHANGES AND CHALLENGES IN THE DESIGN AND IMPLEMENTATION OF A DISTRIBUTED LEARNING ENVIRONMENT

Significant modifications must be made in the instructional design and implementation of courses and in the expectations of instructors and students if a distributed learning environment is to be effective. This section presents a discussion of the nature of some of those modifications and the challenges instructors and students have faced when we've attempted to develop successful courses and programs of study in the DDP.

What will not be successful is a "business as usual" approach in which instructors imagine that they can tweak an existing syllabus to fit the new environment or students assume that they can prepare for the course over the week preceding the class meeting and come prepared to exhibit their understanding during the short synchronous session when the class meets face-to-face. Both instructors and students must adapt to the opportunities and demands of the combination of multiple settings for course activities.

The traditional seminar structure of graduate education frequently entails weekly reading assignments and discussion of the readings during class sessions. A typical syllabus includes an introduction to the scope and content of the course, the objectives or goals of the course, an explanation of class participation standards, papers and projects that each student will be expected to complete, exams that may be required, and a description of the standards of performance for completion or grades. The entire scope of the course is the purview of the instructor, and he or she decides what will be covered and what expectations will be set.

In a distributed learning environment, the instructor may or may not be the sole faculty member responsible for the development of the intellectual activities and interactions that the course may entail. Opportunities and requirements for collaborative planning and implementation of the courses included in the Distributed Doctoral

Program are multiple. Some courses may be taught by one major instructor of record who will have the more traditional responsibilities and control of the course. But other courses may be team-taught or rotate responsibility for activities to different instructors or institutions on a weekly basis. These issues need to be directly addressed at the program level and revisited during the implementation of the program.

Planning course activities for a distributed course is both more time-consuming and more difficult than for a conventional course. Many of the things we take for granted in a traditional classroom do not work without modification in an online course, and some do not work even with modification. We've found that even the most basic class activities often require a major overhaul in order to accomplish the simplest educational objectives. Consider a classroom activity as simple as "spend a few minutes with someone you haven't worked with before and brainstorm/discuss/solve this problem." In this example, the orchestration of hardware, software, and people to support this relatively simple exercise is notable.

Even a course with a single instructor requires modifying instructional design to account for the demands of teaching at a distance and the needs of students taking the course at other institutions. The first few class sessions of a course may allow the instructor to take a measure of the participating students, determining elements of their knowledge and experiences and making modifications to better fit the course to the needs of the students. This is difficult enough in the press of the traditional classroom, but distance and the absence of opportunity for casual interactions before and after each class session make this task even more difficult for the instructor. In addition, many doctoral students have professional obligations that rival those of the faculty. Setting aside time to communicate important background information is difficult. The "salon" aspect of the traditional graduate seminar allows students direct and social interaction with the instructor and with each other. So, students who are new to the environment quickly become acclimated to the culture, including institutional idiosyncrasies and personal quirks of the instructor and fellow students. Informal pre- or post-class or intermission conversations develop the human element of such a course.

The instructor of the course in the distributed learning environment needs to think carefully about how best to facilitate discourse in the synchronous and asynchronous environments of the course. In particular, the instructor must address how best to use time in each aspect of the course. What sorts of discussion questions can best be addressed in the discussion forums of the bulletin board or listserv? What sorts of interactions may reasonably be expected from participants at different sites during chats scheduled outside of the course time? How can the instructor orchestrate the limited synchronous interaction time to encourage interactions from all of the participating students without applying undue pressure? Other issues abound. How can the instructor take into account the fact that some sites may have multiple participants (students and instructors) while other sites may involve only one student? While accountability for grading purposes might be left to participating faculty at the multiple sites, how does the instructor address that issue when he or she is the sole faculty member involved in the course?

Perhaps one of the most perplexing issues for the instructor is how to moderate discussion during the synchronous videoconferencing sessions. The normal social cues that an instructor receives from individuals who wish to contribute to the discussion are not as apparent in the distributed environment. An individual may be monopolizing the discourse because he or she has expertise and experiences that the off-site participants may wish to understand; however, the instructor has an obligation to provide opportunities for all students to engage in the conversations.

In addition, there is a significant danger of cognitive overload during the synchronous sessions, both videoconference and chat. During the pilot course for the Distributed Doctoral Program, one class session had multiple interactions going on in an unstructured manner that resulted in the moderator being bombarded with verbal questions coming over audio, written questions coming from both public and private chat forums, multiple ongoing written private interactions among the participants, and "off-air" oral conversations at the moderator's site. Such information overload quickly shifts the instructor into "survival mode."

The traditional seminar often allows less knowledgeable, beginning students to start interactions with the instructor and "senior" students on a preliminary level, asking for clarification or requesting elaboration and examples. The nature of the discussion forums requires students to develop stances on issues that are public and stay on record for future re-examination. A statement of tentative thinking is difficult to post if the individual later has to defend a position that has changed due to more information. Additionally, a student is pressed to respond quickly to new questions before another participant states his or her position. Agreeing with another's statement has less impact than stating it for the first time, and elaborations, clarifications, and differences are more difficult to develop.

The introduction of asynchronous interactions with students responding to questions and comments posted to a list-serve or to multiple ongoing bulletin-board discussions about significant issues in the course requires continuous effort from the students and the instructor as they struggle to make sense of the discourse that occurs. During the pilot course for the Distributed Doctoral Program, the discussion forum contributions rapidly produced several hundred postings. Keeping track of the issues discussed and the positions taken requires some sort of summative effort from the participants and the instructors of the course. Such summation is arduous and time-consuming. Instructors need to think a priori about the various issues included in the course that may help structure discussion forums.

One model that may serve initially for the development of a distributed learning environment is to think of the course as a "mini-conference." Synchronous, online videoconferencing sessions serve as the featured symposia presentations that may focus on a designated speaker and provide opportunities for participant interactions in a question and answer session. Such a presentation may take up only the beginning part of the course, and breakout sessions may involve either video or chat sessions tailored to aspects of the larger issues that are of interest to participants. Participants might float from one breakout session to another, sampling the threads of the interactions that occur. Breakout sessions might continue throughout the week, either synchronously or asynchronously, as various participants meet to discuss aspects of the ideas and issues under development. Another aspect of this model would allow for sharing of a researcher's work in workshop sessions where the researcher presents applications of research in a multimedia presentation or in shared applications of software or other simulations. This mini-conference model might be expanded to enhance the experiences of students with research being carried on by faculty at diverse institutions without the need to attend large conferences.

One of the biggest challenges for students in a distributed course is the adjustment to this new way of "doing" graduate school. The norms of conventional graduate seminars do not always apply in this part-synchronous/part-asynchronous/part-online/part-offline setting. Rather than requiring a week's preparation for a synchronous "culminating event" as with a conventional graduate seminar, distributed classes tend to be immersive and active all week long. As one student in the pilot DDP course described it, "The structure of the class [has moved] from all the work being on one day to having work spread over the whole week so the content is encountered almost every day of the week." The communication technologies allow anyone to be "in class" at any time on any day. Consequently, distributed courses require students to devote time nearly every day in order to keep up with electronic discussions, multimedia presentations, and other information generated by classmates and faculty.

There is also a fair amount of uncertainty and ambiguity in distributed classes. Most often, when eight or ten graduate students and a faculty member are put in a room together for a three-hour session, they know what to do; this is not the case in distributed doctoral courses. New norms and rules for interacting are needed. It requires adjustments for most graduate students to post their thoughts in a public forum where a text record remains for the class to read and refer back to, to collaborate on research and course work with individuals from other universities and possibly other countries, and to attend class in a virtual classroom in which the professors and students are represented primarily through interactive technology and multimedia. In addition, expectations about feedback and critique on performances both written and verbal that are an important component in the development of professionals need to be publicly and collaboratively developed. Norms must be developed and students must learn to make sense of distributed feedback from learning experiences. Consequently, a high tolerance for ambiguity is helpful in getting through the first few weeks of a distributed course.

BENEFITS OF A DISTRIBUTED LEARNING ENVIRONMENT

Despite the ambiguity and equivocality that tend to accompany distributed courses of the type we've been investigating, students and faculty alike have reported reaping benefits from our pilot efforts in the Distributed Doctoral Program. Students realized that the multiple experiences that their cohorts from other institutions brought to the course added the depth of different perspectives. As one student said, "I am really enjoying reading the comments to questions posted on the web page for our seminar. The past three days I've spent more time than anticipated because the comments are so interesting; I want to read all of them! I am learning a lot about research from the comments by 'classmates.'" A faculty member echoed that sentiment when he said, "The discussion forums have been lengthy and involved deep issues that would not normally be dealt with in the normal development of student researchers. The opportunity to share and contemplate on other perspectives has been useful."

Additionally, students reported benefiting from the appearance of guest scholars in videoconferences who also followed up with students later in the week in chat rooms. Said one student, "The appearances of authors and faculty members was very helpful. The discussion after their presentations was also very good." Another cited the follow-up chat session as being valuable: "I liked having the opportunity to ask [the author] questions in the chat room during the week. It gives you some time to reflect on the information from class."

Perhaps the biggest benefit that both students and faculty reported, however, is the development of professional networks and collaboration partners. Students now look forward to academic conferences as places to catch up with those in their professional network—something that, if students are limited to traditional classroom experiences, usually does not happen until after they have completed graduate school. Distributed courses can put students in touch with scholars who have similar interests. In some cases students have invited faculty met through distributed courses to sit on their dissertation committees. Faculty, as well, have reported enjoying the opportunity to collaborate with students at other institutions. For example, one faculty member noted her interest in a student she met through the course, "[She] and I have formed a great working relationship. We had an excellent online chat on Wednesday and plan to chat approximately every fortnight. She and I are working on similar research topics."

ORGANIZATIONAL ISSUES

A number of administrative and technical issues need to be thought through and sorted out in order to create a distributed learning environment. This section first discusses the administrative overhead required in planning and implementing a distributed course then outlines some technical considerations.

ADMINISTRATION

In developing a distributed course, administrative issues concerning course enrollment and credit, grades, schedules, leadership, and compensation should be worked out at least a full semester prior to the start of the course. First, the course must be "on the books" with the registrar at the appropriate school(s) in time for students to register for the course. The Big Ten's CIC Traveling Scholar program lists its courses in all schools' catalogs and has proven efficient in dealing with this administrative issue, allowing students to cross-enroll at other Big Ten institutions with minimal paperwork. However, programs like this may not be available in every collaborative setting. Accordingly, the course must be listed with enough of the participating schools so that each student who is interested in taking the course may register and receive appropriate credit for the course.

Second, evaluation of student work and the issuing of course grades also should be worked out by the participating faculty members well before the semester's beginning. In general, for pilot courses for the DDP, the participating faculty members at each institution determine course grades for the students enrolled from their school. But, while this has worked well for our needs where faculty members at most of the participating institutions were playing active roles in the course, it may not work as well if there are students involved at institutions without faculty directly participating in the course. At any rate, those who are responsible for evaluation and assessment of students should be identified well in advance of the start of the course.

Third, course schedules, semester start and end dates, and school breaks and holidays vary from one institution to another and have an impact on any synchronous sessions (e.g., videoconferences, etc.) that are planned. In our experience, graduate students are willing to rearrange their schedules to accommodate their learning experiences, but they need to know specific details, such as, for example, when to sign up for the course, that spring break for the course may be a week later, that the course starts and ends a week earlier than the academic schedule at their home institution, or that time zone changes may require mid-course changes in meetings. In our experience, there are likely to be some institutions located in places that do not observe daylight savings, and, the more geographic areas represented in the course, the more difficult these issues are to sort out.

Fourth, while we promote the involvement of faculty at each participating site and any other scholars, experts, and practitioners who may be of value to the students, we also recommend that a course director be identified as a leader and coordinator for each individual course offering. This faculty member acts as an orchestra conductor, ensuring that the requisite pieces keep the same rhythm- and during the times when things get "out of tune" (which will happen no matter how well laid the plans are), the coordinator is the one responsible for restoring harmony.

Fifth, administrative issues concerning which faculty member or members get credit and compensation for teaching the course need to be decided in advance. In the DDP an informal agreement has existed among participating faculty in which a faculty member at one institution assumes the leadership role for a course, receiving compensation for teaching the course, and the others involved offer their expertise and service, knowing that they may be the one teaching the next course. This informal and reciprocal relationship may not work with all institutions, however. In some cases, more formal agreements will be needed in order to facilitate these collaborative relationships.

Those seriously considering offering a distributed class need to be aware of (and have sufficient time and technical support for) the additional day-to-day and week-to-week administrative duties such a course requires, particularly for the director. Maintaining student records, keeping track of faculty participants, staying abreast of the volumes students (and faculty) often contribute to the course, planning the day-to-day course activities, and developing and maintaining a course website consume much more time than do comparable activities in a traditional course. Some faculty have even advocated compensatory course load reductions while teaching at a distance (Porter, 1997).

Keeping track of students enrolled in the course, the universities with which they are affiliated, and whether they are caught up with the work in the course requires diligence and attention to detail. If possible, a teaching assistant otherwise unrelated to the course should be assigned to attend to many of these details. For example, in our current DDP course on research design in mathematics education, several students joined the course during the second week, after it had officially begun. While the students were registered at their home institution, news of their joining the course was slow in reaching the course director. This resulted in those students not receiving course information in a timely fashion and in their not being included in initial mailing lists and directories.

Coordinating the areas of expertise and levels of involvement of the faculty collaborators also requires a time commitment. The course director must orchestrate the talents and knowledge of the participants while juggling the syllabus and course schedule, calling on professional networks to bring in other experts where appropriate. Reminders need to be sent to each faculty collaborator and guest expert several weeks before his or her scheduled "appearance" in the course. Any required technology must be configured and tested to enable each guest to appear in the desired fashion. This often involves coordination with not only the guests, but also with technical support staff at their home institutions.

Graduate students participating in distributed courses have been known to contribute a great deal to threaded discussions, list-servs, and other electronic discussion forums.

The involvement of both students and faculty from multiple institutions results in a more diverse set of ideas and approaches to course material than usually surfaces in a conventional course. That, coupled with the online forums taking the place of most of the conventional "class time," results in complex, often lengthy discussions taking place electronically. This can be beneficial when these discussions produce an audit trail demonstrating how students are making sense of the course material. However, it also may require a time commitment on the part of the instructor that conventional courses do not.

A course website is required in a distributed course. The site is the equivalent of a "homeroom" for students. Any and all information the students need for the course must be posted. The site must be well designed, easy to use, and updated frequently. While creating and maintaining the site is not all that difficult, it does require resources beyond those needed for a traditional course.

TECHNICAL ISSUES

Some thorny technical issues will surface in a distributed course, and many of them seem worse the first time through. For this reason we recommend that faculty who are coordinating distributed courses (and especially those doing so for the first time) tap into the experience of others who have offered similar programs in the past. Even those who are not new at distance education can learn from the experiences of others. Having a resource to draw on when the threaded discussion stagnates, the videoconference question-and-answer session falters, or one of the technologies is not up to its task is invaluable.

All technology used in the distributed course must be thoroughly tested well in advance of the course in circumstances that are as close to reality as possible. However, even under the most rigorous testing conditions, "live" systems may react differently than those tried in testing. Accordingly, coordinators of distributed courses need to be agile and flexible. The extent to which course directors are ready to adapt and accommodate idiosyncrasies in technology, people, or people's interaction with technology directly affects the success of a distributed course.

In the pilot DDP course, we encountered firewalls that would not allow our Australian site to send or receive video, limits on e-mail accounts that, in one case, would not allow the receipt of sizable attachments and, in another, would not allow the sending of sizable attachments, and malfunctioning microphones and other hardware. While these and other issues were eventually resolved, the ability to respond quickly coupled with the patience and goodwill of the collaborators prevented these episodes from becoming disasters.

CONCLUSIONS

In sum, distance education has a great deal to offer doctoral programs in mathematics education. Capitalizing on the core competencies of faculty across institutions provides both graduate students and faculty a number of advantages. Students can be exposed to more ideas and encounter more faculty who are experts and leaders in the field; they can have more opportunities to take highly specialized seminar courses that otherwise would be unavailable; and, they can gain the opportunity to develop professional networks that transcend institutional boundaries early in their careers.

Faculty are able to share their research with a larger audience and work with a greater number of talented graduate students. Regular contact with colleagues at other institutions may set the stage for more collaborative research than would otherwise take place, and faculty members may have more opportunities to teach specialized doctoral-level courses in their areas of interest that they might have inadequate enrollment for at their own universities.

Mathematics education programs can be made stronger through collaborative course offerings; these additional courses and experiences seem to be attractive to highly qualified graduate students; and there is evidence that collaborative opportunities and chances to share one's research through teaching and advising across institutional boundaries helps to retain outstanding faculty members. Therefore, while such programs are not likely to be appealing to all students, professors, and programs, those institutions that do undertake them (and the additional commitments they entail), can expect to share significant rewards.

Information on WebCT is available at http://www.webct.com. Information on CU-SeeMePro and the MeetingPoint conference server is available at http://www.wpine.com

Richard Lesh
School of Education
LAEB 1440 Room 6130
Purdue University
West Lafayette,
rlesh@purdue.edu

Janel Anderson Crider
Department of Rhetoric
University of Minnesota, Twin Cities
64 Classroom Office Building
1994 Buford Avenue
Saint Paul, MN 55108
jcrider@umn.edu

Edith S. Gummer
Department of Science and Mathematics Education
College of Science
Oregon State University
Corvallis, OR 97331
egummer@orst.edu

PART 4: REACTIONS AND REFLECTIONS

CBMS Issues in Mathematics Education
Volume 9, 2001

APPROPRIATE PREPARATION OF DOCTORAL STUDENTS: DILEMMAS FROM A SMALL PROGRAM PERSPECTIVE

Jennifer M. Bay, Kansas State University

In the closing discussion of the doctoral conference, considerable time was dedicated to the question of what constitutes a doctorate in mathematics education. There was also debate as to the need for guidelines or recommendations regarding the nature of doctoral programs in mathematics education. In the case of many Group 3 universities (Reys, et al., this volume) the question is not what defines a *program*, but what defines a *graduate* in mathematics education. Group 3 institutions graduate, at best, about one doctoral student every three years. The NRC reports that since 1980 Kansas State University has granted four Ph.D. degrees with an emphasis in mathematics education. Currently we have two faculty in mathematics education and three part-time doctoral students at various stages in their program. Having so few doctoral students in mathematics education raises many questions, some of which were addressed during the conference. The intent of this paper is to consider the issues of granting Ph.D.s with an emphasis in mathematics education in a Group 3 university setting.

The need for a range of experiences beyond course work was generally accepted as essential in a quality doctoral program in mathematics education. Many such experiences were identified and discussed (Blume, this volume). Some of the opportunities, such as teaching an elementary/secondary methods course, developing a syllabus for this course, and supervising field experience or student teachers, are not as difficult to offer to individual doctoral students. Others, such as engaging in scholarly debate or co-authoring an article for submission to a journal, are more difficult to provide without the collaboration of other mathematics educators (either faculty or other doctoral students). While experienced mathematics educators may write articles and design research individually, doctoral students can gain valuable experience in co-writing and co-participating in research, grant writing, and publishing.

Providing opportunities for other beyond-course experiences is especially problematic for Group 3 schools. For example, working with practicing teachers is an important aspect of a mathematics educator's work and therefore is a valuable experience to include in a doctoral program. It is difficult, however, in a small program, such as the one at Kansas State University, for a doctoral student to plan and facilitate an in-service (e.g., a professional development grant) without extensive participation from the faculty. A challenge for me is how I, as one of two mathematics educators, find time to provide the individual mentorship needed to co-participate in grant writing or in-service while attending to other responsibilities, most of which are devoted to the pre-service preparation of mathematics teachers and master's level students.

The small number of faculty and doctoral students in mathematics education at a Group 3 institution limits focused dialogue and debate about research, issues, or an individual's work. How can Group 3 schools provide opportunities for doctoral students in mathematics education to enter into dialogue with other doctoral students and mathematics educators? In the discussion, one suggestion was to have special sessions or forums scheduled at the regional or annual NCTM meetings to facilitate these discussions. Another solution focused on distance learning. While several people stated that distance learning was not appropriate for all courses and content, we agreed that we have only begun to explore the potential of distance learning in mathematics education. (See Lamb; and Lesh, Crider & Gummer, this volume). The technology for this method of education raises questions about whether there are specific limitations of distance communication. Furthermore, do particular topics or issues lend themselves to this medium? If so, what are they, and how can Group 3 universities facilitate these types of opportunities? Finally, can groups of students from different institutions somehow meet in person to discuss mathematics education?

One conference session focused on core courses in mathematics education (Presmeg & Wagner, this volume). Syllabi were provided that offered insights into what participants' programs required and offered. Other attendees shared their syllabi during the conference. As a mathematics educator in a Group 3 school, I found that examining and reflecting on these syllabi raised a number of issues. For example, at my institution it is impossible to offer courses to doctoral students only, because the overwhelming majority of students in graduate courses are master's candidates. They are more interested in practical teaching ideas than in discussing and/or conducting research. How am I to meet the needs of both master's and doctoral candidates in the same course? My doctoral students will, by necessity, do much of their work in mathematics education through individualized independent study. This raises the second question: What readings, assignments, and activities will most effectively promote learning? Is there a canon of knowledge related to mathematics education that every doctorate in mathematics education should possess? While syllabi distributed at the conference provided ideas for core courses, the content emphasized within those courses warrants further articulation.

Doctoral students typically develop depth in a few research fields and design a dissertation in one of those areas. In larger programs, each doctoral student is able to take courses or do research with faculty who have an interest in the area the student wants to pursue. This is not the case in a Group 3 university. Therefore, another dilemma is whether to allow/encourage my doctoral students to pursue their own research interests or limit their pursuits to my own research interests. That is, should I encourage or even require them to study areas of mathematics education consistent with my own research agenda? One conference attendee said that she does not supervise doctoral dissertations that are out of her own field of research. I have heard this view expressed before. Such a position assumes that entering students know what they want to research, or that there are sufficient faculty to provide a range of choices for research directions. If limiting students to the areas already chosen by faculty is the ideal situation, then should universities state up front that doctoral students will be required to study within these areas?

There was some discussion at the doctoral conference related to designing individual programs based on the career direction (e.g., mathematics coordinator, university teacher educator) each doctoral student expressed when he or she began the program. In my situation, I can create varied experiences for students that will prepare them for some of the positions that were mentioned such as mathematics coordinator and consultant. My current doctoral student, however, wants to join a university faculty. She will not have the advantage of studying with many well-established mathematics educators, and this will likely be a disadvantage to her when she interviews for a position. The challenge I face is how to provide her with broad opportunities to study with other mathematics educators so that she has a more impressive background when she graduates. Perhaps there is a way in which distance learning might be used to fulfill this need.

As we consider the question "What is a doctoral program in mathematics education?" I hope thoughtful attention will be given to the ways in which the common goal might be achieved in programs of various sizes, in particular in the many Group 3 programs around the country that produce only a few doctorates in mathematics education.

Jennifer M. Bay
Kansas State University
251 Bluemont Hall
Manhattan, KS 66506
jbay@ksu.edu

CBMS Issues in Mathematics Education
Volume 9, 2001

PERSPECTIVES FROM A NEWCOMER ON DOCTORAL PROGRAMS IN MATHEMATICS EDUCATION

Alfinio Flores, Arizona State University

Arizona State University established a doctoral program in mathematics education in 1999. Before the conference, I had only a limited vision of doctoral programs guided by my own experience as a student at Ohio State at a time when it was one of the three strongest programs in the country, and my experience in writing the proposal for a new concentration in mathematics education and seeing it become a reality at Arizona State.

The conference expanded my understanding of doctoral programs in many ways and will therefore help our program, too. It helped me realize how much of what I had learned at Ohio State occurred outside the course work. I was also immersed in a rich environment. I was participating in unstructured and informal activities. I was close to first-rate scholars who were very active in the field and to bright, talented fellow students who brought to the program their rich, diverse experiences.

I am now aware of the complexities involved in conducting a successful doctoral program in a sustained way: the diversity of needs of students looking for doctoral degrees, how doctoral programs relate in intricate ways to the institutions in which they are located, and how fragile and dependent on particular people programs can be. Programs that have thrived for years can, in a matter of a few years, nearly cease to exist. It also was striking how very savvy the people in charge of successful programs are about how their own institutions work and about finding external and internal funding sources.

I now view our own program at Arizona State with a new perspective. By hearing about other programs, I realized, for example, that some of the things that I thought were dismal at our institution (such as funding for students) are actually about average and that some of the things we accomplished in an almost natural way (such as collaboration between the mathematics department and the college of education) are not so easily accomplished in other institutions.

Comparing different programs, their goals, and their resources, as well as the different needs and goals of students, I reaffirmed my belief that there is a need for programs that differ in scope, focus, and the type of experiences they offer their students. Looking at the global picture, I see that the program at Arizona State has unique characteristics that can attract students at the national and international levels while the program serves the needs of the local population. Below I discuss some of the characteristics and potential of our doctoral program in light of my experiences at the conference.

INTERDISCIPLINARY EXPERIENCES

The program is a new concentration in an existing interdisciplinary Ph.D. degree program in curriculum and instruction. There are 11 other concentrations. Students in the mathematics education concentration have interdisciplinary experiences of two kinds. First, there are two core courses that students in all concentrations take. These provide students the opportunity to learn about research and curriculum issues in other fields such as science education. Second, students take courses from faculty from the mathematics department and from curriculum and instruction. Students can also choose their cognate areas to provide additional interdisciplinary experiences.

RESEARCH

Doctoral students interested in receiving solid preparation in research can have direct experiences in research projects as well as appropriate courses in different types of methodology in educational research. The four faculty who serve as mentors to students have active programs of research and publication as well as funded grants. Students can be part of an apprentice program in which they are engaged in research and have authentic peripheral participation. The combined areas of expertise of the faculty make it possible to provide a sound core of required research knowledge for students on education topics ranging from those affecting the middle grades to those influencing collegiate mathematics. Students have an opportunity to develop a sense of inquiry and scholarship, address the important ideas in mathematics education, ask meaningful research questions, and participate in a community. The faculty's own work provides examples of important areas of research in mathematics education, such as teaching, fractions, and problem solving. Faculty members have used a variety of methodologies for their research. In addition, faculty and students can participate in an internet-based research methods course with other universities.

EXPERIENCES BEYOND COURSEWORK

It is primarily through the writing of the proposal and the dissertation that each student proves that he or she can conceptualize, conduct, and report research. In addition, through their involvement with funded research, students become acquainted with sources of funding and ways of obtaining research money from them. Faculty regularly encourage and collaborate with graduate students to submit proposals for conference presentations and articles for publication.

Doctoral students can teach undergraduate courses and participate in experiences for in-service teachers to develop their expertise as teacher educators. The department of mathematics offers content courses for elementary teachers and content and methods courses for secondary teachers. The college of education offers methods courses and provides supervised field experiences for prospective elementary and secondary teachers. In addition, faculty members have developed graduate courses in mathematics education for in-service teachers.

Arizona State is well known for the integration of cutting-edge technology into education, and in particular into mathematics. Doctoral students are immersed in an environment rich in technological know-how, advanced courses, and development projects. Three members of the faculty have played leadership roles in introducing technology into the teaching of mathematics at Arizona State.

Faculty involved with the mathematics education program participate actively in the broader scholarly community. They have developed connections to the national and international network of mathematics education researchers and teacher educators. They can assist graduate students in developing their own connections. Faculty have also been actively involved in putting into practice mathematics education reform by changing the way content and methods courses are conducted so that they can serve as models for future teachers.

Reflecting back on the conference and my own experience, I can see that, in addition to what we have already done at Arizona State, we need to acquaint students with several aspects of academic life, such as promotion, tenure, and governance, that figure in career success, and to provide opportunities for them to interact with students from other disciplines on a regular basis.

MATHEMATICAL KNOWLEDGE

The doctoral program at Arizona State offers three different options for students according to their interest: middle school, high school, or college mathematics. The admission requirements in mathematics vary accordingly. For the middle grades, the requirements are flexible enough to allow teachers with K–8 certification (and strong interest and talent in mathematics) to participate in the program. We expect, however, that most students will have a strong undergraduate preparation in mathematics. The entrance requirement for students with interest in secondary mathematics is essentially a major in mathematics, and for those interested in college mathematics, the equivalent of a master's degree in mathematics.

In addition to the strong emphasis on mathematical knowledge for students entering the program, we require doctoral students at Arizona State to develop a better understanding of the mathematics of their level of interest during the program, in both methods and content courses. We want students to exit the program with a profound understanding of the mathematics of their level, and rich pedagogical content knowledge. The program strives to provide students with the appropriate mathematics for their future careers and also to make them capable of communicating about mathematics K–12. The mathematical preparation of those interested in secondary and college-level mathematics will prepare them to teach in mathematics departments.

DIVERSITY OF NEEDS

Conference participants were explicit about the many roles doctoral candidates have to fill. After graduate school, many Ph.D.'s will continue to have more than one role in their careers. It is impossible for one program to be everything to everybody. Nevertheless, the diversity and range of expertise in an institution, including mathematics education faculty as teachers, authors, researchers, program evaluators, directors of funded projects, leaders in professional organizations, as well as other faculty in other departments, such as mathematics, technology and psychology, impact the quality of experiences within a doctoral program in mathematics education. The combined expertise of the faculty guarantees that they can guide students in conducting research in mathematics education, teach an overview of research in mathematics education, conceptualize the K–16 mathematics curriculum, and display a thorough knowledge of the history and evolution of mathematics.

OTHER WAYS IN WHICH THE CONFERENCE WAS HELPFUL

The conference provided many practical ideas. For example, some ideas concerned recruiting students at the local, national, and international levels; others concerned ways to pull together packages of funding to attract the best students. The conference helped me see that the residency requirement should be reconceptualized as a set of needed experiences over several years and as a scholarly, community-building tool, rather than just a requirement of "full-time" studies for a set period of time. The conference also helped make explicit the notion that programs can and should improve over time. They should have goals in terms of development, investment, and consensus. They should also have mechanisms to allow them to learn from experience.

Alfinio Flores
Curriculum and Instruction
Arizona State University
Tempe, AZ 85287-0911
alfinio@asu.edu

WHY I BECAME A DOCTORAL STUDENT IN MATHEMATICS EDUCATION IN THE UNITED STATES

Thomas Lingefjärd, Gothenburg University

At the time I was considering doctoral studies, Sweden did not (and still does not) have any comprehensive programs in mathematics education. Some programs are now developing, but they are nothing like the programs in the United States that have been running for many years. My own attraction to a doctoral program in the United States started in the fall of 1993, when Jeremy Kilpatrick from the University of Georgia was visiting my department at the Gothenburg University as a Fulbright research scholar. In the spring of 1994, I visited the University of Georgia in Athens and became very interested in their doctoral program in mathematics education. The mathematical component of the program was strong; the number of international students was high; and the way the program functioned was very promising in my view.

In the fall of 1994, I was accepted into the program of mathematics education at the University of Georgia. The next year my family and I lived in Athens while I was working on my doctoral program. In the fall of 1996, I returned to Sweden, and since then I have continued my studies at a distance, taking examinations both by e-mail and in Athens. I have traveled back and forth some ten times, usually staying in the United States about two weeks each time. Jeremy has also taught two doctoral courses in Gothenburg, making visits of several weeks' duration and using e-mail exchanges to monitor students' work while he was in America.

While a doctoral student at Georgia, I have more than ever realized that most, maybe all, questions within the field are truly international. Students from Argentina, Colombia, Korea, Sweden, Turkey, the United States, and other countries face similar problems and challenges. These students also bring with them a broad variety of experiences from their home countries, which makes many seminars and discussions rich and interesting.

Moving to a foreign country to pursue a doctorate can be a very challenging experience. While working toward my degree, I played two quite different roles. In Sweden, I was a tenured lecturer and was head of my department at Gothenburg University when I left to attend the University of Georgia. I was later appointed senior lecturer at Gothenburg University while still a graduate student at the University of Georgia. In the hierarchical system that I expect most universities in the United States and Sweden use, it could have been hard for me to survive in both roles simultaneously. But I never hesitated to take on these potentially conflicting roles, primarily because I was encouraged by my major professor. It is a reminder that the influence of a major professor extends far beyond the mentoring done via course work and research. In my case, his support allowed me to complete a doctorate despite playing two roles at once.

One tremendous benefit of working or studying in a country other than your own is that you see your own country or educational system in a new way. Sweden is just about to start doctoral programs in mathematics education. The United States has a century-long history of offering mathematics education as a field of study. Yet it seems that the U.S. debate over the identity of mathematics education and its placement within the university is much the same as in Sweden. At the heart of the debate are the same basic questions: Should mathematics education be a separate department within the university? Should it be a section within a mathematics department, or should it be located within a department of curriculum and instruction? While argument over its organizational structure may continue among universities, placement is ultimately determined by decisions made within each institution.

This conference made it clear to me that the discussion about what constitutes mathematics education and research in mathematics education could continue forever. Perhaps it is necessary for those of us in the community of mathematics education to constantly redefine our field. Mathematics education as a field of study belongs as much within the social sciences as it does within mathematics. It spans the teaching and learning of mathematics at all ages, at all levels, and in all contexts, and, thus, is enormous. The task of deciding where and how such a field of study should be organized in detail is not simply difficult, it is truly impossible. Yet many of the discussions at the conference ended by asking where mathematics education belongs. Is that fruitful? I doubt it.

I consider it more important to establish networks and principles so that the community of mathematics education will open even wider than it is now and happily accept the range of mathematics education: from a one-person community at Stanford University to a large department in the University of Georgia. At Gothenburg, we have an ongoing discussion about where mathematics education should be placed. Should it be within the department of mathematics? Some say to do so would improve the teaching of mathematics in the university classes. Or should it be part of general education? For me it is all a question of environment and milieu—in what environment can our faculty make the greatest contributions?

When I consider the large number of prominent researchers and contributors in our field participating at the conference, some memories strike me as remarkable. There were almost as many opinions about what constitutes research in mathematics education as there were participants. Another issue discussed in some of the seminars was the question of how much mathematics should be studied by those who are themselves mathematics educators. Again there was a wide range of opinion. A third question was whether a doctoral program in mathematics education should be course based or dissertation driven. How many years of residency should be required?

We deal with many of the same questions at Gothenburg. But we still think that much depends on the fact that mathematics education is not established as a research field or a true program of study. Despite the pleasure of meeting and hearing from so many well-known and highly respected people in our field, I left the conference feeling a little disillusioned. The epistemological foundations of the hard sciences are just not possible to construct for our field. Perhaps it is with this uncertainty that our discipline of mathematics education must learn to cope.

Thomas Lingefjärd
Gothenburg University
Department of Education
Section of Mathematics
Box 300
SE 405 30 Gothenburg
Sweden
thomas.lingefjard@ped.gu.se

POLICY—A MISSING BUT IMPORTANT ELEMENT IN PREPARING DOCTORAL STUDENTS

Vena M. Long, University of Tennessee

The National Conference on Doctoral Programs in Mathematics Education generated much thought prior to, during, and after the actual meeting. The preliminary work in gathering background data, identifying the issues, and organizing a program to communicate, interpolate, and extrapolate the information was exemplary. One topic surfaced repeatedly, directly and indirectly, but was never formally or, to my knowledge, informally discussed. The issue is awareness of, knowledge of, and practice in policy analysis, formation, and evaluation.

Education at all levels operates within a political system. In this environment the power lies with those who formulate and those who implement policy. Our educational structure, by default, has concentrated on those who implement policy or, in the case of unsound or unacceptable policy, those who sidestep it. If doctoral students in mathematics education are to become the leaders we envision and need, their preparation must address policy issues. Certainly this training can be through experience just as the methods and techniques of teaching can be learned by doing. This is often akin to trial by fire—as was my introduction to policy issues upon becoming state mathematics consultant for the Missouri Department of Elementary and Secondary Education. However, a methods class can help prepare a teacher to learn faster and more efficiently and introduce future teachers to the resources and support most likely to help them move from novice to master teachers. Likewise, knowledge of theory and practice in the policy arena might prevent the errors that arise from naiveté, thus facilitating sound decision making.

James Fey enumerated why we should have such programs (Fey, this volume). He discussed the need to prepare students to be teacher educators and researchers and the need for roles involving leadership and policy development. In his enumeration of what knowledge, abilities, and dispositions such programs should include, he discussed mathematics and its applicable history and habits of mind, scholarly skills, and disposition for research interpretation, curriculum development and policy analysis. He concluded with proposed methodologies from a generic, rather than a content-specific, perspective.

During the panel presentation "Reflections on the Match between Jobs and Doctoral Programs in Mathematics Education," participants discussed critical, job-related issues that were directly impacted by policy/political issues (Fennell, et al., this volume). Diane Briars, a mathematics consultant for an urban school district; Terry Crites, a department chair; and Susan Gay, a teacher educator who deals with the certification of teachers, must each contend with the ramifications of policy. None of them indicated how, or if, their doctoral programs had prepared them for this challenge. In Arlington this spring, Skip Fennell expressed to me his frustration with the unmathematical nature of the business being discussed by the NCTM Board of Directors during his tenure but, after his term at NSF, he felt much more cognizant of the reality of the politics of mathematics education.

During the conference it was mentioned that a mathematics educator needs a "thick skin." There was discussion about the institutional context of programs and the problems with administration and management that may emerge in that environment as we consider how consortia of departments or institutions might provide richer opportunities for students. People also talked about the relevance or perhaps repetition of experiences within an apprenticeship model and about technology and the policy issues that arise in using technology to teach or do research. Each of these challenges requires us to address, or at least acknowledge, policy.

In discussion generated by Fey's presentation, the need to "connect students with the jobs" seemed a preamble to the panel discussion about what students actually do with doctorates in mathematics education. One of the few points of apparent agreement at the conference was that a program that prepares students to be researchers is not sufficient. A key element of most jobs in academia is successful grant writing. Often the monies that are available are tied to current policy issues. The application of a minimum of knowledge in this area may greatly increase the odds of proposals being funded, thus greatly increasing the odds of our students being successful.

Jeremy Kilpatrick proposed a grid that graphed status against spirit. The goal was to balance the personal needs of the scholar with that of the employee and to balance the needs of academy with that of the marketplace. The integration of policy formation, analysis, and evaluation into mathematics doctoral programs appears to strengthen the programs and the product, creating, if not the desired balance, at least a methodology for working toward such a balance.

Educational decisions at all levels are made through one of three nondiscreet contexts. Most valued by researchers is the theory-based context. Articulating the theory behind what we do and how we do it allows us to generalize to the next decision or dilemma. We work hard to ensure that our decisions are based on sound theory and that our students understand what research has to say about our field and our practice. Yet many practitioners and some politicians decry research and prefer to operate from a historical context. Tradition and common sense, defined as singular personal experience, are the guiding forces. Policy is the context for many other decisions. Teachers teach what will be on the test; they use textbooks from the approved list; and we certify teachers according to state standards.

The degree to which these three contexts diverge on any given issue determines the amount of media hysteria generated. The degree to which these contexts converge determines the successful implementation of any given policy. It is naïve to assume that policy, to be successful, must be grounded in sound theory. Successful politicians will tell you, however, that successful policy is that which can be implemented, that which is doable. The ultimate purpose of any given policy plays second fiddle to the immediate goal of general acceptance. For example, certification of teachers has as its ultimate goal the improvement of student learning through better teachers. However, requiring all teachers to have an acceptable proficiency in mathematics, while theoretically sound, is politically unacceptable.

Jim Hiebert described an improving system as one with a mechanism in place for learning from experience (Hiebert, Kilpatrick & Lindquist, this volume). This mechanism promotes information sharing and the use of information for decision making. Unless we prepare our students for the political realities of education, they cannot become the successful leaders our future demands.

Several avenues exist for introducing policy analysis and evaluation into doctoral programs. These avenues are not mutually exclusive, and some combination might be the best approach. The most obvious vehicle is course work. A course on policy may exist in your institution, but it will probably be in an arena other than mathematics education. Many Ed.D. programs include courses in policy analysis or evaluation so "educational leadership," under whatever name, may be a starting place. Higher education and business schools are the next most likely places. Once course(s) are located, the next task is to determine if the course has the content and the flexibility that will make it worthwhile to mathematics education students. The course should allow students to investigate an area of policy most relevant to them within a context acceptable to their field of study. For example, a policy-analysis course taught from a philosophical context of local autonomy might discredit national standard-setting by a professional organization. While this might be a good exercise in critical thinking, it might not provide the contextual knowledge needed for future issues. A course from a business school might not allow students to consider education as a "business" in selecting project content. While policy analysis and evaluation would seem to dictate an unbiased approach, such is often not the case.

A second avenue exists within existing course work through curricular adjustment. The topic could be added to an existing seminar. Outside resources, such as a professor of such a course from another division, should probably be tapped to provide the theoretical bases. A local school administrator who is literate in mathematics education could discuss implementation of policy issues, and a state department official with similar qualifications could explain the political reality of policy formation and implementation. Since current policy issues are rarely discussed rationally in the press, relying solely on print materials probably would not be sufficient.

Another avenue exists within the experiences we structure for students in doctoral programs. If your state department of education is receptive, doctoral students could serve active or observation-only roles on advisory committees considering such issues as curriculum, testing, and staff development. Doctoral students could job-shadow school district personnel or state department personnel who are involved with appropriate mathematics education issues. State departments might be persuaded that doctoral students are an excellent source of inexpensive brain power. Doctoral fellowships underwritten by state government could provide great experience for students and increase the people power of chronically understaffed departments of education. Such experiences would often be more varied, less repetitive, and more substantive than observational experiences in the classroom or in administration.

In less structured ways we can ensure some level of awareness of policy. Whatever we experience as mathematics educators, whether at the national, state, local, or institution level, our doctoral students should have a place "looking over our shoulders." As their research questions develop, we should be asking them to consider the policy implications inherent in their topics. Whether oriented to curriculum, instruction methodology, or assessment, policy can impede or impel movement toward good, research-supported practice.

At the doctoral level, the future of our profession can be greatly enhanced by producing students who are cognizant of the role politics plays in education, who understand the need to formulate sound policy in the political arena, and who know the process by which policy becomes practice. I can attest that firsthand experience with policy analysis, formation, implementation, and evaluation produces a "thick skin." Mathematics education must emphasize appropriate and significant mathematics curriculum, instruction and assessment; appropriate and significant mathematics; and the pertinent theories of learning, techniques of research, and the philosophical and historical aspects of education. To the massive challenge of incorporating all of this, at least an introduction to policy issues should be added.

Vena Long
317 Claxton Addition
University of Tennessee
Knoxville, TN 37996
vlong@utkux.utcc.utk.edu

CBMS Issues in Mathematics Education
Volume 9, 2001

My Doctoral Program in Mathematics Education— A Graduate Student's Perspective

Gay A. Ragan, Southwest Missouri State University

The National Conference on Doctoral Programs in Mathematics Education was certainly a unique and unforgettable experience. While I was a conference participant, I also was one of the few people who got the opportunity to view the conference through other perspectives: that of a doctoral student and of a working member of the Doctoral Conference Project staff.

At the time of the conference, I was a fourth-year mathematics education doctoral student at the University of Missouri. I had completed my comprehensive examinations and started writing my dissertation proposal. During the conference discussions I found that, as a graduate student, I "compared and contrasted" my doctoral program at the University of Missouri with what I learned of other institutions' programs.

Based on my status as a mathematics education graduate student at the University of Missouri (and having an office next door to Robert Reys), I, along with another graduate student, Bob Glasgow, was invited in the fall of 1998 to be a working member of the conference staff. One goal of the project was to determine the status of U.S. doctoral programs in mathematics education prior to the national conference. In order to do so, Bob Glasgow and I worked with Dr. Reys and the organizing committee to develop a comprehensive survey of doctoral programs in mathematics education, collect and analyze the survey data from U.S. institutions, and prepare a status report (Reys, et al., this volume). As a result of my work on the survey, I found my conference perspective unique because I had learned a bit about the faculty, students, and graduate programs of all institutions represented prior to the conference.

Based upon my view of the conference through a split lens, I reflected on my own doctoral program in mathematics education at the University of Missouri. This paper shares some reflections about the "core components" and the "beyond coursework" experiences of my doctoral program.

Core Components

Preparation in Research

Lester and Carpenter summarized the conference discussions about the research preparation of doctoral students (this volume). In their closing comments, they acknowledged that doctoral students' preparation in research varies dramatically among institutions, and go on to state:

> At some—perhaps many—institutions, the only direct involvement in
> research required of doctoral students is the dissertation. At others—
> perhaps only a few—doctoral students grapple with research problems
> and issues throughout their programs of studies, and the dissertation
> stage serves as an indicator of the students' readiness to enter the world
> of research on their own (p. 63).

From my perspective as a graduate student (and borrowing a slogan from the
United States Marine Corps), "We are the few, the proud, …the graduate students in
mathematics education at the University of Missouri." My preparation in research was
enhanced by a program requirement to be continually enrolled in mathematics education
seminar courses (one three-hour-credit course each semester) until completion of the
comprehensive examinations. These seminars addressed a variety of themes that enabled
me to build a "core" knowledge base of research in mathematics education. Through
these seminars, I eventually found an area of special research interest and began to
build a knowledge base related to the specific topic of my dissertation research. Through
coursework designed to provide essential knowledge of research issues, methods, trends,
and results in the field of mathematics education, I was able to study and, more
importantly, to participate in quantitative and qualitative research projects. Thus, my
preparation in research was as both a participator and as a spectator.

Due to my status as a full-time graduate student and my need for financial support,
I have had other opportunities to work as a graduate research assistant for externally
funded projects. I have been involved in conducting research with faculty and have
worked as a part of research teams made up of other graduate students, thus being a part
of a "community of researchers" (in press). I would agree (and what doctoral student
would not) with Lester and Lambdin (as cited in Lester and Carpenter, this volume)
that my dissertation research is a valuable experience. Besides gaining experience as the
person with the sole responsibility for research design, data collection, data analysis,
and reporting of the results, the dissertation research is serving as a personal indicator
of my own readiness to enter the world of research and succeed on my own. I feel the
knowledge base from formal course work together with the opportunities to participate in
a range of research projects during my doctoral program will serve me well in my career
as both a consumer and producer of research in mathematics education.

PREPARATION FOR MATHEMATICS.

In contrast to the recommendations made by Dossey and Lappan (this volume) the
mathematics preparation in my program extends across all levels: elementary, middle,
and secondary. While the doctoral program typically requires a master's degree in
mathematics, the content is not specifically identified and is typically tailored to the
career direction of each candidate. In reflecting on my personal needs and experiences,
if I had been asked at the start of my program to choose one level of mathematics
specialization, I would have had difficulty choosing between the secondary level and the
collegiate level (advanced topics) since I had spent equal amounts of time teaching at
those levels. Furthermore, as I worked as a graduate assistant, I became involved in the
Show-Me Project at the University of Missouri. This project's focus is the dissemination
of information about the five National Science Foundation-funded standards-based
middle grades curricula. Thus, it has been important that I have mathematics preparation
relevant to the middle grades level. Most recently, I have taken advantage of an
opportunity to participate in a research project focused on mathematics at the elementary

level. I am grateful that my doctoral program has helped me to understand better the mathematical demands at different levels. Upon entering my doctoral program, I could not have accurately predicted the various levels of mathematics I would be involved in nor could I have predicted a single area of specialization for my mathematics preparation. Therefore, based upon my experiences, I am skeptical of specific recommendations for the mathematics preparation (Dossey and Lappan, this volume) made at the conference.

PREPARATION IN MATHEMATICS EDUCATION

At the University of Missouri, there is no set of "core" courses required for preparation in mathematics education, but any graduate student in our program can identify a set of "core" courses. This "core" set would focus on the following topics, which are offered regularly as mathematics education seminars: history of mathematics curriculum, number sense, learning theories, manipulatives and/or technology, and problem solving. Such course offerings are based on our faculty's areas of expertise. Other seminar courses offered are often based on students' interests or suggestions. During my program, there were four mathematics education faculty and less than a dozen doctoral students. Therefore, only a limited number of courses could be offered, and master's level students were often enrolled in them.

In reflecting upon the discussion at the conference about preparation in mathematics education, I believe the program at the University of Missouri is "in-line" with what most other U.S. institutions require. While unofficially there does not exist a set of "core" courses at the University of Missouri, each doctoral student takes approximately 30 credit hours in mathematics education. I along with my fellow graduate students recognize that to become well-prepared mathematics educators, we must take it upon ourselves to read and learn about areas outside those covered in such a set of required "core" courses. While I support the idea of a set of "core" courses for everyone, programs at different institutions should also take advantage of the expertise of their faculties and the interests of their students when designing programs to prepare graduates in mathematics education.

PREPARATION FOR COLLEGIATE TEACHING

Discussion at the conference about the preparation of graduates for collegiate teaching caused many participants to reflect on the opportunities provided within their programs. Questions arose such as "Can a graduate student ever receive enough preparation for collegiate teaching during their program?" and "Should the graduate program be designed to include the preparation of teaching mathematics courses as well as mathematics education courses?"

Personally, one opportunity I wanted from my graduate program was to teach a methods course for preservice mathematics teachers. During my doctoral program I was given the opportunity to teach a methods course, and to supervise student teachers in mathematics at both the middle grade and secondary level. While both of these experiences were invaluable, I realize that I have only begun to gain the knowledge and experience I need to be prepared for these roles.

On the other side of the coin, most graduates at the University of Missouri gain good preparation for collegiate teaching of mathematics because they either get the opportunity to teach as graduate teaching assistants for the mathematics department or, as in my case, enter the program with collegiate teaching experience in mathematics.

In reflecting on the conference discussions about preparation for collegiate teaching, the overarching question was "How can it be done?" I would first say it must be recognized that the preparation needed to design and teach undergraduate (and graduate) courses in mathematics education is different from the preparation needed to teach collegiate mathematics. The University of Missouri program tries to address preparation in both these teaching areas by providing opportunities for graduate students to teach in the mathematics department and in mathematics education. Prior to teaching a mathematics education course, graduate students are asked to "sit-in" or assist a faculty member in teaching such a course. Since most of the teaching opportunities are with undergraduate students, we are required to take a graduate-level seminar course each semester until completion of the comprehensive examinations (and are often encouraged to audit such seminar courses while in the dissertation stage). Being enrolled enables us to continually build an extensive knowledge base of mathematics education issues. Furthermore, I found these seminar courses serve as a model in preparing us to teach such graduate-level courses in the future.

Finally, I encourage all programs to provide graduates with opportunities to work in the field with student teachers, their partner teachers, and local school districts. One of the foundations of being prepared to teach at the collegiate level, whether it be mathematics or mathematics education courses, is to keep in touch with the reality of what is happening in schools. While it is very important that preservice teachers know what could (and should) be happening in the teaching and learning of mathematics, it is of equal importance to such students that we, as their (future) professors, recognize what is actually happening in schools related to the teaching and learning of mathematics. Opportunities to teach mathematics and mathematics education courses, learn from professors as models in seminar courses, and work with undergraduates and teachers in local public schools have prepared me well to teach at the collegiate level.

BEYOND COURSEWORK

Of all the variations in U.S. doctoral programs in mathematics education identified by the survey (Reys, et al., this volume) there was very little difference in program requirements involving comprehensive examinations (written and oral) and the dissertation component. Graduate students at the University of Missouri are required to successfully complete written comprehensive examinations in both mathematics and mathematics education. After a student's program committee has reviewed the examinations, an oral examination is conducted. The dissertation component of the program at the University of Missouri follows the "traditional" format used by most U.S. institutions surveyed. However, conference discussions of alternative formats for the dissertation component, such as reporting the research in a form suitable for publication, were well received by the faculty at the University of Missouri. While I cannot speak for other graduate students either at the University of Missouri or at other institutions, I can say, from my own experience, that these two "hurdles" heightened my level of anxiety about graduate studies. Yet, at the same time "clearing" these "hurdles" provided me with a sense of professional success. To me, each of these components, in its own unique way, required me to be able to see, understand, and explain the "big picture" in terms of mathematics education.

Before reflecting further on my personal experiences beyond coursework at the University of Missouri, I believe I must provide a background of my progression as a graduate student. When I entered the graduate program, I had collegiate teaching experience in mathematics but no public school teaching experience. The faculty member who interviewed me prior to my applying to the graduate program (who later became my graduate program advisor) stressed the importance of gaining public school teaching experience before completing a doctoral degree in mathematics education at the University of Missouri. I took her advice and taught for three years in a local public high school. Her advice was exactly right. I now recognize that the courses in the program were much more meaningful to me as a result of my public school teaching experience. In addition, this experience proved valuable to me when teaching and supervising preservice teachers. At the same time, I began the doctoral program as a part-time graduate student, taking one mathematics education seminar course in the evening each semester and playing "catch-up" during the summer by taking mathematics and mathematics education courses. Next, my advisor told me about opportunities that would be valuable to me but would require me to be a full-time graduate student. Again, her advice was good. By becoming a full-time graduate student, I was able to take multiple courses and seriously concentrate on issues in mathematics education that led to my area of research interest and my dissertation topic. The move also allowed me to shift from thinking only like a practitioner and begin examining issues as a researcher and teacher educator. As a full-time graduate student, I was able to work closely with the faculty members and become involved in projects that have led to research and grant-writing opportunities. In addition, my full-time status allowed me to collaborate on projects with other graduate students, which led to opportunities to write for publications and conduct national conference presentations.

CONCLUSION

I am grateful for the opportunity to be a part of the Doctoral Conference Project staff and participate in the national conference. This experience provoked me into stepping back and reflecting upon my own doctoral program. While I have done a general self-reflection of my doctoral program at the University of Missouri, the best reflection of my preparation came during a recent interview by a search committee chairman. In reviewing my curriculum vitae, he said, "You sure have been busy doing lots of things since beginning your doctoral program." I took this as a compliment and as a reflection of the opportunities I experienced. It also demonstrates that I have attended a high-quality doctoral program that offers a breadth of experience and depth of preparation in both the areas of core components and beyond coursework. In my judgment, these components should be prominent in all doctoral programs in mathematics education.

Gay A. Ragan
School of Teacher Education
Southwest Missouri State University
Springfield MO 65804
gar098f@smsu.edu

PART 5: IDEAS FOR ACTION

CBMS Issues in Mathematics Education
Volume 9, 2001

IMPROVING U.S. DOCTORAL PROGRAMS
IN MATHEMATICS EDUCATION

James Hiebert, University of Delaware
Jeremy Kilpatrick, University of Georgia
Mary M. Lindquist, Columbus State University

Two lessons can be learned from reflecting on sessions and conversations at the conference and by reading the papers in this volume: First, there is a shared interest in searching for ways to improve the doctoral programs in mathematics education in the United States, and second, there is considerable diversity along many dimensions of this issue that pose challenges for collective, systematic efforts to improve. In this paper, we offer a framework for improving complex systems like doctoral education, review several challenges that confront improvement efforts, and suggest next steps that might be taken within a larger improvement plan. Because the mathematics education community is just at the beginning of a conscious collaborative process to reconsider and improve doctoral education, we present our comments as discussion points rather than formal recommendations.

IMPROVING THE FUNCTIONING OF COMPLEX SYSTEMS

One way of conceptualizing the challenge of improving mathematics education doctoral programs is to think of mathematics education, including the training of its leaders, as a complex system. Improving the training process means improving the system rather than changing individual features. The advantage of this perspective is that it focuses attention on continuous, cumulative efforts that will help all programs improve rather than on sporadic efforts that often are unevenly distributed and temporary.

There are no recipes for improving complex systems, but there are a few guidelines that can be abstracted from analyzing such processes at work in a variety of professional and learning settings (Siegler, 1996; Simon, 1996; Wilson & Daviss, 1994). Readers are encouraged to think about the mathematics education doctoral system in the United States as they read the four general guidelines outlined below.

ASSESS INITIAL CONDITIONS

A first step toward improvement is to know the starting point. What are the conditions of the system now? Without good information about the current conditions, it is impossible to know what the next steps should be. Gathering such information often requires collecting data from participants at different points in the system. The information of most use is the status of the system at each point with respect to the remaining three guidelines.

SET GOALS

In order to improve, the participants in a system must be relatively clear about the direction in which they would like to go. Without goals in some form, it is impossible to know whether change in the system represents progress or just change. Clearly stated goals also provide the guidance needed to make principled decisions within the system, to marshal resources to promote improvement, to communicate with those outside the system, and to evaluate the success of new, innovative programs.

If the whole system is to improve, consensus must be reached about the goals. Participants must share a commitment to achieve them. Without shared goals, the efforts in one part of the system cannot be used to improve the system as a whole. If the system does not yet have clearly expressed goals, considerable effort often is needed to reach a consensus on goals, but the investment in this process pays off substantially.

The goals themselves can be expressed in many different forms including, on the one hand, ideals that capture the vision of what the system ultimately would like to be or, on the other hand, minimal standards for effective functioning. Regardless of the form, it is important that the goals be expressed clearly enough so that participants can tell whether the system is making progress toward reaching them.

Goals can change. Conditions inside and outside the system often change, which prompts a need to reassess and revise the goals. This reconsideration is a natural part of the improvement process.

DEVELOP PLANS FOR MOVING FROM INITIAL CONDITIONS TO GOALS

A third guideline is that some ideas must be proposed for how the system can move from its starting point toward the goals. In the world of complex systems, well-formed theories for improvement are not likely to be available, especially ones that prescribe an algorithm for achieving goals. But some proposals must be offered that describe the mechanisms or processes that will move the system forward.

Just as the goals can change along the way, the processes that guide the improvement activities are likely to change. In fact, the system needs a steady diet of new ideas to find new, more effective, ways to move toward its goals.

DOCUMENT AND SHARE INFORMATION ABOUT IMPROVEMENT EFFORTS

Steady, continuous improvement requires that the system learn from its experience. To prevent the system from constantly starting over, mechanisms must be designed to record the experiences of participants at each point in the system and share this information throughout the system. In this way, the whole system can take advantage of what is learned at one point and build, in a cumulative way. Value is placed on even small achievements as long as they can be shared with other participants committed to reaching the same goals.

SPECIAL CHALLENGES FACING COLLABORATIVE IMPROVEMENT EFFORTS

Imagine the collection of U.S. doctoral programs in mathematics education as a complex system. Where does it stand vis-à-vis the four guidelines proposed above? Because mathematics educators are just beginning to attend directly to improving the system of doctoral education in the United States, it is not surprising that there is little collaborative activity that addresses any of the four guidelines. Before outlining some next steps that might jump-start a collective improvement process, we identify two features of the current system that pose the most significant challenges.

ABSENCE OF STANDARDS

The absence of system-wide standards for doctoral programs is, perhaps, the most serious challenge facing systemic improvement efforts. Shared standards have never existed for U.S. programs in mathematics education (see Donoghue, this volume, for an historical account). Indeed, participants in the system have grown accustomed to creating their own standards at each local site. Developing a consensus on goals or standards is a significant step because it will require a change in practice. It will remove some of the isolation and autonomy of individual programs in favor of a shared commitment to improving the system of doctoral education. Changing practice in this way involves changing culture, and cultural changes are neither quick nor easy.

Currently, the standards of individual programs can vary widely. This creates a problem for system-wide improvement because if different programs are working toward different goals or standards, then what is learned at one site is of little use at other sites. Individual programs can improve, but the system as a whole has no way of improving. More precisely, the system has no way to know whether it is improving.

The absence of standards poses a further problem for new programs that are being developed. No guidelines are available that might help developers design programs that would ensure that graduates have the competencies expected by the profession. New programs can copy older established ones, but that is a poor model for building new programs. Those doing the copying will have little or no way of knowing whether the features they are copying are related to program quality and to the success of graduates. (For additional observations on problems of setting up a new program, see Thornton, et al., & Wolff, this volume.)

A consequence of the absence of standards is the huge disparity among the requirements of different programs (Reys, et al., this volume). For example, the number of semester hours in the programs ranges from 45 to 125. Furthermore, the programs are vastly different in size. At least 115 institutions have awarded doctoral degrees in mathematics education during the past 18 years, with more than one third of these institutions graduating only one or two students during this time. It is reasonable to conclude that students in these different programs are having very different experiences (see Fennell, et al., this volume, for comments confirming that diversity). But this runs counter to the notion that there might be some common knowledge or experiences needed by all graduates who wish to enter leadership positions in mathematics education.

The issue of diversity defines the environment in which mathematics education doctoral programs are situated. We have just argued that some aspects of this diversity are probable consequences of the absence of shared standards and might be reduced in an improved system. But that is not the whole story.

DIVERSE CONSTITUENTS

Diversity in advanced training in mathematics education presents both problems and opportunities for improvement. Diversity of doctoral programs is not merely the unintended outcome of programs that happen to have been constructed in different ways; it reflects the diversity within mathematics education as a professional field.

Mathematics educators play a variety of professional roles and engage in a variety of professional activities. They engage in, among other activities, research in teaching and learning mathematics, teacher education, curriculum development and supervision, policy analysis, collegiate mathematics instruction, and instructional leadership in schools (see the papers on core components in this volume for a discussion of program components that address some of these activities). The activities of mathematics educators require a variety of skills. It is not surprising that a variety of training programs have emerged.

In addition, students enter doctoral programs with various backgrounds and different skills. Some have advanced training in mathematics; others have advanced training in education or psychology. Some have pre-college teaching experience; some have research experience; some have worked in business and industry; and some are highly skilled technologically. This diversity, coupled with different professional goals and interests, suggests that multiple programs are needed to provide students with the expertise they will need. (Some ways in which programs are addressing this diversity at a time of declining enrollments are illustrated by Aichele, et al., this volume).

As we have noted, the diversity in programs poses challenges for efforts to develop a system-wide set of standards or goals. The process of setting common standards must be sensitive to the diversity inherent in the field. But that diversity also presents opportunities. It provides a rich and continuing resource for new program ideas. Programs that differ significantly from the norm could serve as local experiments, with the results shared throughout the system. The diversity that currently exists could provide, in part, the steady diet of new ideas needed to improve the system over time.

WHAT ARE THE NEXT STEPS?

ASSESS CURRENT CONDITIONS

Presuming that the mathematics education community would like to launch a long-term continuous improvement effort for its doctoral programs, what steps should be taken first? Following the framework outlined earlier, the first step should be an assessment of current conditions. Such an assessment matches well the view of the conference participants. In a conference follow-up survey in which all of the participants responded, 96% supported the need to assess current conditions of doctoral education in the U.S.

The two questions to be resolved to enable this data collection process are the following: What information should be collected? Who should collect it? There was considerable agreement on the follow-up survey about the kind of information to collect. Over 90% of the respondents voted to collect data from each program on these features: the number of mathematics education faculty; the number of graduates each year; where the graduates found jobs; the number of fulltime students; and the number of part-time students. Additional information surely will be useful and can be added to the list as the process gets underway.

The question of who will collect the information is more difficult to resolve. A wide range of suggestions were offered during the closing session of the conference including current or ad hoc committees or officers of the National Council of Teachers of Mathematics, the Mathematical Association of America, the Special Interest Group for Research in Mathematics Education of the American Educational Research Association, or a special agency formed for this purpose. Reys et al., this volume) offer some specific suggestions that should be considered.

Given that data collection will be the first step in a process that could lead to system-wide improvement efforts, we recommend that the task be viewed as more than technical. Whatever agency becomes the data collector, some mathematics educators who are informed about the short- and long-term goals of the effort must be involved. Decisions will need to be made early on about the exact nature of the questions, the form in which the data will be most useful, and the policies and procedures used to collect and disseminate the information. Moreover, changes will be needed in all of these aspects as the process moves forward. Awareness of how other academic disciplines and professions collect and disseminate this kind of information will be helpful.

DEVELOP CONSENSUS ON PROGRAM GOALS

Formulating a shared set of standards or goals for doctoral training in mathematics education is the second step in an improvement effort. This likely will be the most challenging step in the process but, for reasons outlined earlier, we believe it is necessary for system-wide improvement.

The first challenge is to agree that setting goals is itself an important part of the process. When asked on the follow-up survey if they felt that general guidelines for 'core' components for doctoral programs in mathematics education should be developed, 62% of the conference participants said yes, 25% said no, and 13% were unsure. The primary concerns of those who said no were that common standards might eliminate diversity among programs and stifle creativity in developing innovative programs. As noted earlier, goal-setting efforts can and must be sensitive to these concerns.

We recommend a general strategy that includes taking stock of the goals that currently exist within individual programs, initiating a planned discussion leading to consensus of broad goals, and then gradually moving toward increased specificity as the field evolves. A first step should be to include on the data collection instrument questions regarding each program's goals. It is likely that current goals come in many shapes and sizes. The questions should allow respondents to express their goals in whatever form currently is being used.

The second step should begin the process of moving from individual program goals to a shared set of goals. To achieve consensus, these probably will be expressed initially at a general level. They might take one or more of several forms: statements of value; descriptions of exiting expertise; minimal standards of performance; core program components; recommended practices; or others. Questions on the data collection instrument could assess whether there is an emerging consensus about the form in which the shared goals should be developed.

The goal-setting process should be conducted in a fully open, inclusive environment but also in a deliberate, time-sensitive way. It will be important to solicit comments from those inside and outside of mathematics education. The standards-setting process used recently by the National Council of Teachers of Mathematics (NCTM) is one model that might be emulated. Goal-setting processes used in other academic and professional fields also should be investigated.

The third step should be to continue the discussion in order to increase specificity and clarity of goals wherever possible. The clearer and more specific the goals, the more useful they are for assessing quality of existing programs, measuring the success of improvement efforts, and guiding the development of new programs.

It is likely that the development of goals, both in their broad and specific forms, will proceed unevenly in different areas. For example, among those conference participants who agreed that guidelines should be set for core components of programs, there already was substantial agreement in some areas, including mathematics, mathematics education, research expertise, and professional experiences (e.g., teaching). This agreement means that the process might move forward quite quickly on some fronts. We recommend that areas of consensus be pursued and formalized into shared goals while areas of disagreement continue to be discussed. There is no need to wait for all matters to be resolved before at least some system-wide goals are formed. Setting goals is a process to be continued throughout the life of the system.

A major issue is who will initiate, monitor, and coordinate the goal-setting process. One option would be for an organization such as NCTM, the National Academy of Sciences, or the Conference Board of the Mathematical Sciences to convene a group to get the process underway. Another would be for senior faculty from the pool considered for the invitational conference to organize a specially funded project.

DEVELOP A COLLECTIVE THEORY OF IMPROVEMENT

A theory of how U.S. doctoral programs in mathematics education improve is not a prerequisite for improvement, but it will be an outcome of collaborating on improving the system. The form that such a theory takes will be created as the community engages in deliberate and focused conversations about these issues.

In a real sense, a theory for program improvement provides a target to shoot for as individual improvement efforts are documented and shared across the system. Ideally, such information will contain reasoned hypotheses about why changes would improve programs, examples of such changes, and the outcomes of the changes. These local theories with examples and outcome data can then be synthesized into more comprehensive hypotheses and theories about how the improvement process is working.

It is likely that this grand-sounding theory formation effort will begin by examining mechanisms that already have been initiated or easily could be imagined. For example, both the conference on which this volume is based and the volume itself represent improvement mechanisms that might be significant features of the process. One way of measuring their effect would be to document activities spawned by the conference. A joint effort by the University of Maryland, the University of Delaware, and the Pennsylvania State University to develop a collaborative doctoral program is one example. Others surely could be recorded and tracked. A second kind of mechanism, suggested at the conference, is the creation of a website that would serve as a clearinghouse for information as well as an interactive communication center for system improvement projects. If implemented, this website and its surrounding activity should be studied to understand its role in the improvement process.

Information useful for constructing theories of improvement is not limited to activities in the United States. Advanced training in mathematics education around the world provides a rich source of information which should be used throughout the process (Bishop, this volume).

SUCCESSFUL IMPROVEMENT IS BECOMING HIGH STAKES

There are good reasons to believe that improving the "average" doctoral program in mathematics education is critical for securing the future of mathematics education as a field of study. That is, system-wide improvement should not be viewed just as a laudable goal; it should be viewed as imperative.

IMMEDIATE DEMANDS FOR NEW LEADERS

Over the next few years, universities will be searching for an increasing number of doctoral graduates in mathematics education. According to Reys (2000), 51% of university faculty in mathematics education will be eligible for retirement within two years, and 78% will be eligible for retirement within ten years. It is likely that the demand will be even greater than these figures indicate because of the additional emphasis and resources targeted for improving the training of mathematics (and science) teachers. It is difficult for graduates to make up for weak training once they are on the job, so it is crucial that graduate programs provide strong exiting expertise. The increasing demand for trained mathematics educators represents a tremendous opportunity, and responsibility, to move the field forward with highly qualified graduates.

LEADING THE PROFESSION IN AN AGE OF ACCOUNTABILITY

The trend toward increased accountability for the effectiveness of educational practice cannot and should not be ignored. Classroom teachers across the country are facing this issue on a daily basis. Educational training at the highest level should take the lead in accepting and embracing this responsibility by developing high quality programs that are accountable (for further discussion of this issue, see Fey, this volume).

As doctoral programs move toward accountability, they need to address the tensions between being a scholar and being an employee that will exist in the careers their graduates pursue. Programs need not only to prepare doctoral students to do scholarly work but also to function well as employees; accountability measures will need to address both. Moreover, there are tensions between the academy and the marketplace in every career employing doctoral graduates in mathematics education. Programs need to be accountable for preparing students adequately to resolve these tensions.

Improving complex systems is a continuing process that yields small changes over time. But those changes can accumulate to yield lasting and fundamental improvements rather than quick and temporary fixes. We believe it is important for the mathematics education community to take the initiative and begin a rational long-term process of improving its programs for training coming generations of doctoral students.

James Hiebert
School of Education
University of Delaware
Newark, DE 19716
hiebert@udel.edu

Jeremy Kilpatrick
105 Aderhold Hall
University of Georgia
Athens, GA 30602-7124
jkilpat@coe.uga.edu

Mary M. Lindquist
Columbus State University
4225 University Avenue
Columbus, GA 31907-5645
lindquist_mary@colstate.edu

REFERENCES

REFERENCES

Adams, J. A. (1902). *Correlation between mathematics and physics in American high schools.* Unpublished master's thesis, Teachers College, Columbia University, New York.

AMTE News (1999). Announcement of National Conference of Issues Related to Doctoral Programs in Mathematics Education. *8*(1), 8.

Archibald, R.C. (1938). *A semicentennial history of the American Mathematical Society, 1888–1938.* New York: American Mathematical Society.

Batanero, M. C., Godino, J. D., Steiner, H. G. & Wenzelburger, E. (1992). The training of researchers in mathematics education—Results from a survey. *Educational Studies in Mathematics, 26,* 95–102.

Bednar, C. (1910). *Educational grounds for unified mathematics.* Unpublished master's thesis, University of Chicago.

Bishop, A.J. (1992). International perspectives on research in mathematics education. In D.A.Grouws (Ed.) *Handbook of research on mathematics teaching and learning.* (pp. 710–723) New York: Macmillan

Bishop, A.J., Clements, M.A.,Keitel, C., Kilpatrick, J. & Laborde, C. (Eds.). (1996). *International handbook of mathematics education.* Dordrecht, Holland: Kluwer

Brown, J. F. (1911). *The training of teachers for secondary schools in Germany and the United States.* New York: Macmillan.

Central Association of Science and Mathematics Teachers. (1903). *School Science, 3,* 113–114.

Central Association of Science and Mathematics Teachers. (1950). *A half century of science and mathematics teaching.* Oak Park, IL: Author.

Committee on Undergraduate Science Education, Center for Science, Mathematics and Engineering Education, National Research Council (1999). *Transforming undergraduate education in science, mathematics, engineering, and technology.* Washington, DC: National Academy Press.

Day, R. P. (1999). *Course Syllabus for Math 305.* Normal, Illinois: Illinois State University. [http://www2.math.ilstu.edu/~day/courses/old/305/www.html]

Donoghue, E. F. (1998). In search of mathematical treasures: David Eugene Smith and George Arthur Plimpton. *Historia Mathematica, 25,* 359–365.

Donoghue, E. F. (in press). The emergence of a profession: Mathematics education in the United States, 1890–1920. In G.M.A. Stanic & J. Kilpatrick (Eds.), *A history of school mathematics*. Reston, VA: National Council of Teachers of Mathematics.

Duke, N. K., & Beck, S. W. (1999). Education should consider alternative formats for the dissertation. *Educational Researcher, 28* (3), 31–36.

Erwin, W. C. (1910). *Study of a preparatory school class in algebra.* Unpublished master's thesis, University of Chicago.

Explanation. (1905). *School Science and Mathematics, 5,* 133.

Gibb, E. G., Karnes, H. T., & Wren, F. L. (1970). The education of teachers of mathematics. In P. S. Jones & A. F. Coxford, Jr. (Eds.) *A history of mathematics education in the United States and Canada* (Thirty-second yearbook, pp. 299–350). Washington, DC: National Council of Teachers of Mathematics.

Gilliland, A. M. (1903). *Mathematics in the great public schools of England.* Unpublished master's thesis, Teachers College, Columbia University, New York.

Glasgow, R. (2000). *An investigation of recent graduates of doctoral programs in mathematics education.* Unpublished doctoral dissertation, University of Missoui.

Goodlad, J. I. (1994). *Educational renewal: better teachers, better schools.* San Francisco, CA: Jossey-Bass Publishers.

Grouws, D. A. (Ed.). (1992). *Handbook of research on mathematics teaching and learning.* Reston, VA: National Council of Teachers of Mathematics.

Holzinger, K. J. (1922). *The indexing of a mental characteristic.* Unpublished doctoral dissertation, University of Chicago.

Information on WebCT is available at http://www.webct.com/

Information on CU-SeeMePro and the Meeting-Point conference server is available at http://www.wpine.com/

International Commission on the Teaching of Mathematics. (1911). *Training of teachers of elementary and secondary mathematics.* The American report, committee no. 5. U.S. Bureau of Education Bulletin 1911, no. 12. Washington, DC: Government Printing Office.

Jackson, L. L. (1906). *The educational significance of sixteenth century arithmetic from the point of view of the present time* (Doctoral dissertation, Columbia University). New York: Author.

Jones, P. S. & Coxford, A. F. Jr. (1970). Mathematics in the evolving schools. In P. S. Jones & A. F. Coxford, Jr. (Eds.) *A history of mathematics education in the United States and Canada* (Thirty-second yearbook, pp. 9–89). Washington, DC: National Council of Teachers of Mathematics.

Journal for Research in Mathematics Education (1999). *Announcement of National Conference, 30*(1), 118.

Kelly, A. and Lesh, R. (2000). *The Handbook of Research Design in Mathematics and Science Education.* Mahwah, NJ: Lawrence Erlbaum.

Kilpatrick, J. (1992). A history of research in mathematics education. In D. A. Grouws (Ed.), *Handbook of research on mathematics teaching and learning,* (pp. 3–38). New York: Macmillan.

Klein, F. (1932). *Elementary mathematics from an advanced standpoint.* Translated from 3rd German edition by E.R. Hedrick and C.A. Noble. New York: Dover Publications.

Klein, F. (1893). Inaugural address. New York Mathematical Society *Bulletin, 3,* 1–3.

Klein, F. (1911). *The Evanston colloquium, lectures on mathematics.* New York: American Mathematical Society. (Originally published by Macmillan, 1893)

Krathwohl, D. (1994). A slice of advice. *Educational Researcher, 23*(1), 29–32, 42.

Lester, F. K. & Lambdin, D. V. (in press). From amateur to professional: The emergence and maturation of the U.S. mathematics education research community.. In G.M. A. Stanic & J. Kilpatrick (Eds.), *A history of school mathematics.* Reston, VA: National Council of Teachers of Mathematics.

Lindquist, T. (1911). *Mathematics for freshmen students of engineering.* Doctoral dissertation, Univesity of Chicago.

Long, R. S., Meltzer, N. S. & Hilton, P. J. (1970). Research in mathematics education. *Educational Studies in Mathematics, 2,* 446–468.

Luckey, G. W. A. (1903). *The professional training of secondary teachers in the United States.* New York: Macmillan.

Mathematical supplement of School Science. (1903a). *School Science, 2,* n.p. (Announcement following p. 428)

Mathematical supplement of School Science. (1903b). *School Science, 2,* n.p. (Announcement following p. 486)

McIntosh, J. A. & Crosswhite, F. J. (1973). *A survey of doctoral programs in mathematics education.* Columbus, OH: ERIC Information Analysis Center for Science, Mathematics and Environmental Education. (ERIC Document Reproduction Service No. ED 091 250).

Michigan State Normal School. (1894–1895). *Yearbook.* Ypsilanti, MI: Author. (Held in Eastern Michigan University archives)

Monroe, W. S. (1915). *A history of arithmetic in the United States, with emphasis upon the influence of Warren Colburn in directing the development of arithmetic as a school subject.* Doctoral dissertation, University of Chicago. (Reprinted as *Development of arithmetic as a school subject,* 1917, in U.S. Bureau of Education Bulletin no. 10).

Moore, E. H. (1926). *On the foundations of mathematics* (First yearbook of the National Council of Teachers of Mathematics). New York: Bureau of Publications, Teachers College, Columbia University. (Reprinted from *Science, 17* (1903), 401–416).

National Educational Association. (1893). *Report of the committee on secondary school studies* [Committee of Ten]. (U.S. Bureau of Education whole no. 205). Washington, DC: U.S. Government Printing Office.

National Educational Association. (1970a). Report of the committee of the Chicago section of the American Mathematical Society. In J. K. Bidwell & R. G. Clason (Eds.), *Readings in the history of mathematics education* (pp. 195–209). Washington, DC: National Council of Teachers of Mathematics. (Reprinted from *Journal of proceedings and addresses of the thirty-eighth annual meeting*, 1899, Chicago: Author)

National Educational Association. (1970b). Report of the mathematics conference to the committee on secondary school studies [Committee of Ten]. In J. K. Bidwell & R. G. Clason (Eds.), *Readings in the History of Mathematics Education* (pp. 129–141). Washington, DC: National Council of Teachers of Mathematics. (Reprinted from *Report of the committee on secondary school studies*, U.S. Bureau of Education whole no. 205, 1893, Washington, DC: U.S. Government Printing Office)

National Research Council. (1998). *Summary Report 1996: Doctoral Recipients from United States Universities.* Washington, D. C.: National Academy Press.

National Research Council. (1999). *Summary Report 1997: Doctoral Recipients from United States Universities.* Washington, D. C.: National Academy Press.

NCTM Newsletter (1999). Announcement of National Conference of Issues Related to Doctoral Programs in Mathematics Education. *35*(6), 7, January. Reston, VA: National Council of Teachers of Mathematics.

New York Mathematical Society. (1892). *New York Mathematical Society list of members, constitution, by-laws: June, 1892.* New York: Author.

Notes. (1892). New York Mathematical Society *Bulletin, 1,* 80,124.

Osborne, A. R., & Crosswhite, F. (1970). Forces and issues related to curriculum and instruction, 7–12. In P. S. Jones & A. F. Coxford, Jr. (Eds.) *A history of mathematics education in the United States and Canada* (Thirty-second yearbook, pp. 153–297). Washington, DC: National Council of Teachers of Mathematics.

Parshall, K. H., & Rowe, D. E. (1994). *The emergence of the American mathematical research community, 1876–1900: J. J. Sylvester, Felix Klein, and E. H. Moore.* Providence, RI: American Mathematical Society and London Mathematical Society.

Perry, John. (1901). The teaching of mathematics. In J. Perry (Ed.), *Discussions on the teaching of mathematics*, British Association meeting at Glasgow (pp. 1–32). New York: Macmillan.

Porter, L. (1997). *Creating the Virtual Classroom: Distance Learning with the Internet.* New York: Wiley

Report of the American Commissioners of the International Commission on the Teaching of Mathematics. (1912). U.S. Bureau of Education Bulletin 1912, no. 14. Washington, DC: Government Printing Office.

Reys, R. E. (2000). Doctorates in Mathematics Education: An Acute Shortage, *Notices of the American Mathematical Society, 47*(10), 1267–1270.

Richardson, J. E. (1903). *Influence of Pope Sylvester II (Gerbert) upon mathematics about the year one thousand.* Unpublished master's thesis, Teachers College, Columbia University, New York.

Rudolph, F. (1962). *The American college and university: A history.* New York: Random House/Vintage books.

Russell, J. E. (1900a). The function of the university in the training of teachers. *Teachers College Record 1,* 1–11.

Russell, J. E. (1900b). The organization and administration of Teachers College. *Teachers College Record 1,* 36–59.

Russell, J. E. (1937). *Founding Teachers College.* New York: Bureau of Publications, Teachers College, Columbia University.

Russell, J. E. (various dates). *Papers.* Milbank Memorial Library, Teachers College, Columbia University.

Schoenfeld, A. H. (1999). The core, the canon, and the development of research skills: Issues in the preparation of education researchers. In. E. Lagemann & L. Shulman (Eds.), *Issues in Education Research: Problems and Possibilities,* pp. 166–202. New York: Jossey-Bass.

Schreiber, E. W., & Warner, G. W. (1950). Central association and the journal. In *A half century of science and mathematics teaching* (pp. 1–38). Oak Park, IL: Central Association of Science and Mathematics Teachers.

Shulman, L. S. (1986). Those Who Understand: Knowledge growth in teaching. *Educational Researcher, 15* (2), 4–14.

Sierpinska, A. & Kilpatrick, J. (Eds.). (1998). *Mathematics education as a research domain: A search for identify.* Dordrecht, The Netherlands: Kluwer.

Siegler, R. S. (1996). *Emerging minds: The process of change in children's thinking.* New York: Oxford University Press.

Silver, E. A. & Kilpatrick, J. (1994). E pluribus unum: Challenges of diversity in the future of mathematics education research. *Journal for Research in Mathematics Education, 25,* 734–754.

Simon, H. A. (1996). *The sciences of the artificial (3rd ed.).* Cambridge, MA: MIT Press.

Smith, D. E. (1900). The teaching of elementary mathematics. New York: Macmillan.

Smith, D. E. (1905). Réforms á accomplir dans l'enseignement des mathématiques. *L'Enseignement Mathématique 7,* 469–71.

Smith, D. E. (various dates). Professional papers. Rare Book and Manuscript Library, Columbia University.

Smith, D. E., & Ginsburg, J. (1934). *A history of mathematics in America before 1900* (Carus monograph no. 5). Chicago: Mathematical Association of America.

Stamper, A. W. (1909). *A history of the teaching of elementary geometry, with reference to present-day problems* (Doctoral dissertation 1906, Columbia University). New York: Teachers College, Columbia University.

Stanic, G. M. A. (1986). The growing crisis in mathematics education in the early twentieth century. *Journal for Research in Mathematics Education, 17*(3), 190–205.

Stanic, G. M. A. (1987). Mathematics education in the United States at the beginning of the twentieth century. In T. S. Popkewitz (Ed.), *The formation of school subjects* (pp. 145–175). New York: Falmer.

Stone, C. W. (1908). *Arithmetical abilities and some factors determining them* (Doctoral dissertation, Columbia University). New York: Teachers College, Columbia University.

Syracuse University. (1896–1897). *Annual.* Syracuse, NY: Author.

Teachers College, Columbia University. (1902–1903). *Announcement.* New York: Author.

Teachers College, Columbia University. (1895–1896). *Circular of information.* New York: Author.

Teachers College, Columbia University. Department of Mathematics. (1906–1907). *Courses for the training of teachers of mathematics and other information relating to the department.* New York: Author.

Teachers College, Columbia University. Department of Mathematics. (1911–1912). *Courses for the training of teachers of mathematics and other information relating to the department.* New York: Author.

Teachers College, Columbia University. *Dictionary Catalogue* (vol. 32). New York: Author.

Teachers College, Columbia University. (1902). *Report of the Dean.* New York: Author.

The meetings of the association. (1904). Association of Teachers of Mathematics in the Middle States and Maryland *Bulletin, 1,* 18–19.

Tyack, D. (Ed.). (1967). *Turning points in American educational history.* Waltham, MA: Ginn/Blaisdell.

University of Chicago. (1894–1895, 1902–1903, 1904–1905, 1906–1907, 1909–1910, 1910–1911, 1911–1912). *Annual register.* Chicago: Author.

University of Chicago. (1900–1920). *Convocation programs.* Chicago: Author.

University of Michigan. (1892–1893). *Calendar.* Ann Arbor, MI: Author.

Upton, C. B. (1907). *Modern calculating machinery and its bearing on the teaching of mathematics.* Unpublished master's thesis, Teachers College, Columbia University.

Valero, P. & Vithal, R. (1998). Research methods from the "north" revisited from the "south". In A. Olivier & K. Newstead (Eds.) *Proceedings of the 22nd conference of the International Study Group for the Psychology of Mathematics Education,* vol 4 (pp. 153–160), Stellenbosch, South Africa: University of Stellenbosch.

Walbesser, H. H. & Eisenberg. T. (1971). What research competencies for the mathematics educator? *American Mathematical Monthly, 58,* 667–673.

West, M. (1912). *The differential and integral calculus in secondary schools.* Unpublished master's thesis, Teachers College, Columbia University, New York.

Wilson, K. G., & Daviss, B. (1994). *Redesigning education*. New York: Holt.

Young, J. W. A. (1900). *Teaching of mathematics in the higher schools of Prussia*. New York: Author.

Young, J. W. A., & Jackson, L. L. (1904). *Arithmetic*. New York: Appleton.

APPENDICES

APPENDIX A. LIST OF PARTICIPANTS

Name	University	Email Address
Aichele, Douglas	Oklahoma State University	aichele@math.okstate.edu
Arvold, Bridget	University of Illinois	arvold@uiuc.edu
Barger, Rita	University of Missouri-Kansas City	bargerr@umkc.edu
Barnes, David	University of Missouri-Columbia	barnesd@missouri.edu
Battista, Michael	Kent State University	MTBattista@aol.com
Bay, Jennifer	Kansas State University	jbay@ksu.edu
Becker, Jerry	Southern Illinois University	jbecker@siu.edu
Beem, John	University of Missouri-Columbia	mathjkb@showme.missouri.edu
Bernhardt, Robert	East Carolina University	bernhardtr@mail.ecu.edu
Bishop, Alan	Monash University, Clayton, Australia	alan.bishop@education.monash.edu.au
Blume, Glen	Pennsylvania State University	bti@psu.edu
Boaler, Jo	Stanford University	joboaler@stanford.edu
Briars, Diane	Pittsburgh Public Schools	briars@pps.pgh.pa.us
Bright, George	National Science Foundation	gbright@nsf.gov
Carpenter, Thomas	University of Wisconsin	tpcarpen@facstaff.wisc.edu
Confrey, Jere	University of Texas	jere@mail.utexas.edu
Crites, Terry	Northern Arizona University	terry.crites@nau.edu
Crosswhite, Joe	Ohio State University (Professor Emeritus)	Phone: 417-886-9182
Donoghue, Eileen	Columbia University-Teachers College	efd3@columbia.edu
Dossey, John	Illinois State University	jdossey@math.ilstu.edu
Erickson, Dianne	Oregon State University	ericksod@ucs.orst.edu
Fennell, Skip	Western Maryland College	ffennell@wmdc.edu
Fey, James	University of Maryland	jf7@umail.umd.edu
Fleener, Jayne	University of Oklahoma	fleener@ou.edu
Flores, Alfinio	Arizona State University	alfinio@asu.edu
Gay, Susan	University of Kansas	sgay@ukans.edu
Greenes, Carole	Boston University	cgreenes@bu.edu
Grouws, Doug	University of Iowa	douglas-grouws@uiowa.edu
Heid, Kathleen	Pennsylvania State University	ik8@psu.edu
Hiebert, James	University of Delaware	hiebert@udel.edu
Holmquist, Mikael	Gothenburg University, Sweden	mikael.holmquist@ped.gu.se
Hunting, Robert	East Carolina University	huntingr@mail.ecu.edu
Johnson, Martin	University of Maryland	mj13@umail.umd.edu
Kilpatrick, Jeremy	University of Georgia	jkilpat@coe.uga.edu
Lamb, Charles	Texas A & M University	celamb@acs.tamu.edu

(continued)

173

Appendix A. (continued)

Name	University	Email Address
Lambdin, Diana	Indiana University	lambdin@indiana.edu
Lappan, Glenda	Michigan State University	lappan@msu.edu
Lester, Frank	Indiana University	lester@indiana.edu
Lewis, Jim	University of Nebraska	jlewis@math.unl.edu
Lindquist, Mary	Columbus State University	lindquist_mary@colstate.edu
Lingefjard, Thomas	Gothenburg University, Sweden	thomas.lingefjard@ped.gu.se
Long, Vena	University of Tennessee	vlong@utk.edu
Maher, Carolyn	Rutgers University	cmaher@rci.rutgers.edu
Matthews, Pamela	Montgomery College, Germantown	pamjr@worldnet.att.net
Owens, Doug	Ohio State University	owens.93@osu.edu
Post, Thomas	University of Minnesota	postX001@tc.umn.edu
Presmeg, Norma	Illinois State University	npresmeg@msn.com
Reys, Barbara	University of Missouri-Columbia	reysb@missouri.edu
Reys, Robert	University of Missouri-Columbia	reysr@missouri.edu
Richardson, Lloyd	University of Missouri-St. Louis	lloyd_richardson@umsl.edu
Rock, David	University of Mississippi	rockd@olemiss.edu
Schultz, Karen	Georgia State University	kschultz@gsu.edu
Shaughnessy, Mike	Portland State University	mike@mth.pdx.edu
Sherman, Helene	University of Missouri-St. Louis	helene_sherman@umsl.edu
Sowder, Judith	San Diego State University	jsowder@sciences.sdsu.edu
Spikell, Mark	George Mason University	mspikell@gmu.edu
Stiff, Lee	North Carolina State University	lee_stiff@ncsu.edu
Thompson, Pat	Vanderbilt University	pat.thompson@vanderbilt.edu
Thornton, Carol	Illinois State University	thornton@math.ilstu.edu
Tunis, Harry	National Council of Teachers of Mathematics	htunis@nctm.org
Wagner, Sigrid	Ohio State University	wagner.112@osu.edu
Wearne, Diana	University of Delaware	wearne@udel.edu
Weinstein, Gideon	U.S. Military Academy	gideonw@banet.net
Whitenack, Joy	University of Missouri-Columbia	whitenackj@missouri.edu
Wilson, James	University of Georgia	jwilson@coe.uga.edu
Wolff, Kenneth	Montclair State University	wolffk@mail.montclair.edu
Graduate Students **Name**	**University**	**Email Address**
Chavez, Oscar	University of Missouri-Columbia	oc918@mizzou.edu
Fields, Wanda	University of Missouri-Columbia	wmf0a8@mizzou.edu
Glasgow, Bob	University of Missouri-Columbia	bglasgow@sbuniv.edu
Kim, Ok-Kyeong	University of Missouri-Columbia	ok182@mizzou.edu
Ragan, Gay	University of Missouri-Columbia	gar098f@smsu.edu
Simms, Ken	University of Missouri-Columbia	ksimms@hhs.columbia.k12.mo.us
Steencken, Elena	Rutgers University	steencke@rci.rutgers.edu
Taylor, Mark	University of Missouri-Columbia	pmta43@mizzou.edu
Wasman, Dee	University of Missouri-Columbia	wasmandg@appstate.edu

Appendix B. Conference Agenda

Conference on Doctoral Programs in Mathematics Education
October 16–19, 1999

Saturday, October 16

5:00–5:30	Registration
5:30–6:30	Social Hour
6:30–7:30	Dinner
7:30–8:00	Overview of Conference *Robert Reys*
8:00–9:00	Evolution of Mathematics Education as a Discipline and Doctoral Programs *Eileen Donoghue*

Sunday, October 17

7:00–8:00	Breakfast
8:15–9:00	Doctoral Programs in Mathematics Education: Results from a Survey *Robert Reys*
9:00–10:00	Major Components of Doctoral Programs in Mathematics Education *Jim Fey*
10:00–10:30	Break
10:30–11:45	Breakout Groups: Round One (Choose One)
	Preparation in Research: What preparation is needed? *Tom Carpenter & Frank Lester*
	Preparation in Mathematics: What and for whom? *John Dossey & Glenda Lappan*
	Preparation in Mathematics Education: Is there a basic core for everyone? *Norma Presmeg & Sigrid Wagner*
	Preparation for Collegiate Teaching: How can it be done? *Diana Lambdin & Jim Wilson*
12:00–1:00	Lunch
1:15–2:30	Breakout Groups: Round Two (Choose One)
	Preparation in Research: What preparation is needed? *Tom Carpenter & Frank Lester*
	Preparation in Mathematics: What and for whom? *John Dossey & Glenda Lappan*
	Preparation in Mathematics Education: Is there a basic core for everyone? *Norma Presmeg & Sigrid Wagner*
	Preparation for Collegiate Teaching: How can it be done? *Diana Lambdin & Jim Wilson*
2:30–3:00	Break

(continued)

175

Appendix B. (continued)

3:00–4:15	PANEL:
	Reflections on the match between jobs and doctoral programs in mathematics education *Diane Briars; Terry Crites; Skip Fennell, Facilitator; Susan Gay; Harry Tunis*
5:00–6:00	SOCIAL HOUR
	DINNER (ON YOUR OWN)

MONDAY, OCTOBER 18

7:00–8:00	BREAKFAST
8:15–9:45	Reports from Breakout Groups on Preparation *Joe Crosswhite, Facilitator*
9:45–10:15	BREAK
10:15–11:00	DISCUSSION SESSIONS (CHOOSE ONE)
	Beyond Coursework Dissertations Distance Learning
11:00–*11:45*	*DISCUSS*ION SESSIONS (CHOOSE ONE)
	Beyond Coursework Dissertations Recruiting and Funding Doctoral Students
12:00–1:00	LUNCH
1:15–2:15	Challenges and *Dilemmas in Developing Doctoral* Programs *Pat Thompson*
2:15–2:45	*BREAK*
2:45–4:00	*SIMULTANEOU*S PANEL PRESENTATIONS/DISCUSSIONS (CHOOS*E ONE*)
	Organizing a new doctoral program in mathematics education *Robert Hu*nting; Mike Shaughnessy; Judith Sowder; Carol Thornton, Facilitator; Kenneth Wolff
	Reorganizing exis*ting doctor*al programs Doug Aichele, Facilitator; Jo Boaler; Carolyn Maher; David Rock; Mark Spi*kell*
5:30–6:30	*SOCIAL HOUR*
6:30–7:30	DINNER
7:30–8:30	International perspectives *of doctoral programs* *Alan Bishop*

TUESDAY, OCTOBER 19

7:00–8:00	BREAKFAST
8:15–9:30	Reports from Group Discussions Martin Johnson, Facilitator
9:30–10:00	BREAK
10:00–11:30	Where does this lead? Jim Hiebert; Jeremy Kilpatrick, Facilitator; Mary Lindquist